Symptoms, Investigations and Treatment in Eating Disorders

Symptoms, Investigations and Treatment in Eating Disorders

A quick reference guide for all ED professionals

Dr. Murali K Sekar MBBS DPM MRCPsych
Consultant Psychiatrist in Eating Disorders
Community Eating Disorder Service
Hertfordshire, UK

Dr. Krishnakumar Muthu MBBS MD
Assistant Professor in Internal Medicine
University of South Dakota/ Sanford School of Medicine
Avera McKennan Hospital and University Health Centre
South Dakota, USA

AuthorHouse™ UK Ltd.
1663 Liberty Drive
Bloomington, IN 47403 USA
www.authorhouse.co.uk
Phone: 0800.197.4150

© 2013 by Dr. Murali Sekar, Dr. Krishnakumar Muthu. All rights reserved.

No part of this book may be reproduced, stored in a retrieval system, or transmitted by any means without the written permission of the author.

Published by AuthorHouse   07/02/2013

ISBN: 978-1-4817-9810-5 (sc)
ISBN: 978-1-4817-9811-2 (e)

Any people depicted in stock imagery provided by Thinkstock are models, and such images are being used for illustrative purposes only. Certain stock imagery © Thinkstock.

Because of the dynamic nature of the Internet, any web addresses or links contained in this book may have changed since publication and may no longer be valid. The views expressed in this work are solely those of the author and do not necessarily reflect the views of the publisher, and the publisher hereby disclaims any responsibility for them.

Contents

Symptoms and Signs 1

Minerals and Trace elements.. 35

Vitamins 85

Water Soluble Vitamins.. 89

Fat Soluble Vitamins 117

Renal Function Tests 133

Liver Function Tests 159

Endocrinology... 173

Fluid Balance 203

Miscellaneous Topics 211

Glossary of Terms and Definitions................................ 225

Further reading list 230

Index... 231

To Thilaga, Shruthi and Shree

To my parents

To my teachers and patients (MKS)

To my Family, Friends and Patients who have been a constant source of inspiration in my quest of knowledge (KM)

Acknowledgements

I would like to offer my sincere thanks to Prof. Bob Palmer for his advice and support. This book would not look like anything what it is without his input. I would like to thank my teachers from Leicestershire Eating Disorders team for encouraging me to complete this project. I would also like to thank the team members from Eating Disorders Unit in Cheshire and Wirral Partnership NHS from where I got the idea of writing this book. My sincere thanks to staff members of my current team (Community Eating Disorders Team, Hertforshire Partnership NHS Foundation Trust) who offered valuable suggestions and help which went a long way in completing this book.

Finally I would like to thank staff from AuthorHouse publishers (Jean Swan, Mars Alma, Stacy Canon and Kathy Lorenzo) for their wonderful guidance and prompts which helped me to finish the book.

MKS

Preface

Physical health monitoring and risks related to this aspect of eating disorders is always a complex and often daunting task for any professional working in this field. We often come across a situation when a patient asks us about her/his blood test results. He/she would like to know what is sodium, for example, and what does it do in the body. They also would like to know how low their sodium level is and if it is marginal what should they do to improve it. Often our colleagues who come from a non medical professional background e.g psychology, nursing etc come across similar situations.

This book is an attempt to make us equipped to deal such situations. In this book, we have included some of the common topics as well as some specific tests we come across at times in this field. We have also included some common clinical problems of eating disorders e.g laxative abuse and their management.

We have tried our best to avoid using medical jargons as much as possible. We hope that most of us despite the professional background would have some level of familiarity with these terms. We hope that this book will be a good pocket reference guide. The reader will get basic information about what they need to know and do about common clinical situations in eating disorders setting. We have used a shade coding system as much as possible in the management section of various topics.

All we did in this book is to review some of the common medical facts through an eating disorder prism and alter the facts to meet the need of a clinician working in the field of eating disorders.

Disclaimer

The shade-based coding system used in this book is for guidelines only. The aim is to give an idea about symptoms that are associated with a reference range for a clinical test. We cannot predict that this is always the case. In clinical practice onset of symptoms depend upon a number of other factors.

The treatments suggested are for practical guidelines only. Individual specialist centres and medical wards may vary in their practice. In case of doubt, experts in the field should be consulted.

Normal values given in this book could differ from some other resources. In case of doubt, please check with the local laboratory where the test was done.

The Shade Code:

This shade means either the clinical picture is commonly seen when the patient is in the community or the patient can be treated in the community. The observed alteration in the blood test is marginal.

This shade means either the clinical picture is commonly seen when the patient is in the eating disorders unit or that the patient needs to be treated in the EDU. The observed alteration in the blood test is moderate.

This shade means either the clinical picture is commonly seen when the patient is in the medical ward or the patient needs to be treated in the medical ward. The observed alteration in the blood test is serious

Symptoms and Signs

Symptoms and signs

Eating disorders affect all the systems in the body. Hence the resulting symptoms differ in its origin. Differentiating symptoms with regard to its origin are a challenging task to a clinician treating this group of patients. Since eating disorders are often treated by professionals like psychologists, nurses their expertise in identifying physical symptoms and recognising when to seek medical help could vary. Nevertheless this is an essential skill since often they will be the frontline and sometimes sole professionals treating the patients.

This chapter will provide some introduction to symptoms and signs the ED professionals need to be familiar. We have done the best to help the reader to identify symptoms and signs that are urgent requiring immediate attention from those that are not. As we have attempted throughout this book, the content of this chapter would be useful for professionals with various levels of medical knowledge. In case of doubt, we would recommend the reader to seek opinion and advice from experts in the field.

Subsection for each system in the body i.e gastrointestinal, cardiovascular etc could be found in this chapter. The signs and symptoms are arranged alphabetically with, wherever possible, an indicator of severity of that symptom (what a patient reports) or sign (what a clinician identifies). Only the symptoms and signs that are relevant to ED are included in this chapter. Obviously, patients with an eating disorder could develop other medical conditions (e.g malignancy, autoimmune disorders). The symptoms of those conditions are not included in this book. History from the patient, relatives and other professionals involved in the patient care would be immensely helpful in such cases.

Dr. Murali Sekar, Dr. Krishnakumar Muthu

General Symptoms and signs:

Finger nail abnormality:

Clubbing: This is characterised by loss of angle between the nail and finger. Nail bed would feel soft. There is an increased curvature of the nails both longitudinally and horizontally. Clubbing is mainly due chronic lung conditions like bronchiectasis, lung abscess and COPD. Other conditions like heart disease, cirrhosis of liver could also lead to clubbing of fingers. This condition is not directly related to ED conditions, but indirect links to chronic nutritional deficiency e.g increased chance for lung infections) could result in aforementioned medical conditions that can in turn lead to clubbing.

Koilonychia: spoon shaped nails usually due to iron deficiency anaemia (occasionally due to ischaemic heart disease) which is due to poor intake of food that are rich in iron. Diagnosis confirmed by low haemoglobin and ferritin level.

Onycholysis: thickened, dystrophic nail, which is usually, separate from the nail bed. This condition is seen in hyperthyroidism. Hence this may be seen in patients with thyrotoxicosis factitia (intake of thyroid supplement to lose weight). Thyrotoxicosis is confirmed by thyroid hormone level estimation. Onycholysis may also be seen in psoriasis. This condition is not directly linked to eating disorders.

Beau's lines on nails: These are the transverse furrows in the nails. This is a result of any long term illness hence could be seen in chronic eating disorders with nutritional deficiency.

Onychomedesis: frequent shedding of nails which could be a manifestation of chronic nutritional deficiency.

Mee's lines: Single white transverse lines or bands on the nails. Chronic renal or heart failure can present with this. These conditions could be seen in chronic ED patients with impairment of functions of these organs.

Yellow nails: could be due to low albumin level (a form of protein). Hypoalbuminaemia is commonly seen in low weight anorexia nervosa patients.

Hand Joints:

Primary Osteoarthrosis: Prolonged menopause leads to this clinical condition. Characterised by paired bony nodes (Heberden's nodes) on terminal interphalangeal joints. Can also affect other bigger joints like elbow joint. Usually require treatment with NSAIDs for pain and swelling.

Diffuse hair loss: seen in iron deficiency anaemia or chronic severe nutritional deficiency. Commonly seen in ED clinical setting. Also, common is lanugo hair, a fine hair replacing the adult type course thick hairs.

Finger nails are one of the sites to be examined for anaemia. Pale subconjunctiva, hands, face are the other indicators of anaemia. Size of the red blood cells (normal size is between 76-96 fl) helps to identify the cause of anaemia. In cases of iron deficiency, the cell size would be less than 76 fl. Vitamin B12 and/or folic acid deficiency causes megaloblastic anaemia in which cells will be more than 96 fl. If anaemia is due to any chronic condition like chronic renal failure, the cells size could be preserved (normocytic anaemia).

Hypothermia:

Hypothermia is clinically diagnosed when the body temperature is less than 35 degrees Celsius (95 degree Fahrenheit). When temperature is low, we need to double check the temperature from another site, to confirm the reading as the actual temperature could be still lower. Rectal thermometers confirm hypothermia. Severe hypothermia is defined as temperature less than 33 degrees (91 degree Fahrenheit). In anorexia nervosa, due to lowering of the basal metabolic rate, the temperature could be low. These patients are also prone to hypothermia when exposed to cold weather. To add to the complexity, these patients are not able to identify the symptoms of hypothermia. Hence wearing protective clothes against body heat loss is one of the important psychoeducation these patients need.

Common causes for hypothermia in anorexia are low metabolic rate, poor body fat content to insulate against heat loss or protection against low environmental temperature.

Treatment: Rewarming at a rate of 0.5 degrees/hour is usually the first step in the management of hypothermia. Faster rate could be fatal. Other measures like ventilation, Intravenous access, catheterization may be necessary depending upon the clinical need.

Mouth Lesions:

Poor nutrition, recurrent vomiting (alteration of oral bacterial content due to stomach acid), erosion of teeth enamel (leading dental caries), low levels of immunity, nutritional deficiency can affect the health of the mouth. Infections of gum due to herpes simplex and fungal infection (candidiasis) could be present in ED patients. These conditions, if they occur, can be difficult to treat. Deficiency of vitamin B12, riboflavin and nicotinic acid can cause atrophic glossitis (atrophy of papillae of the conditions respond to treatment with replacing the deficient vitamin.

Parotid Swelling:

Confined to the sides of the face but can extend to involve other salivary glands. Swelling and pain (if infected) can be the presenting symptom. Rinsing the mouth with salt water or lemon flavoured liquid would be effective in reducing the swelling.

Thyroid swelling: In the anterior part of the neck, check for any diffuse or localised swelling in thyroid glands. History of use of thyroid supplement could be useful as is the history/ blood tests of altered thyroid function.

Gynaecomastia:

Presence of firm disc of tissue underneath the nipple is essential to the diagnosis of gynaecomastia. If it is only fat as seen in obese individuals, the term gynaecomastia should not be used. In theory, low levels of testosterone could produce gynaecomastia. Other hormones like oestrogen are also low in low weight individuals along with low fatty tissue. Hence men with anorexia nervosa do not have gynaecomastia.

Lanugo Hair:

Fine hair distributed widely in the body. Lanugo hairs are seen in anorexia nervosa but not in malnutrition due to other conditions.

Spider Naevi:

Spider naevi are very small (size of a pin head), red lesions with radiating blood vessels. These are usually seen in the anterior chest wall and abdomen. When pressure is applied on these pin head centres, the lesion tend to blanch. In patients who take oestrogen, these benign lesions could appear. Reduction in dose or stopping Oestrogen makes the lesions disappear. Hence when patients receive oral contraceptive pills (OCP) for osteoporosis, clinician could use this to check the patient's compliance with OCP.

Spider naevi can be seen in normal conditions. They will be usually less than 3 in number. Other conditions like liver failure, pregnancy are also known to cause spider naevi.

Purpura:

Purpura is from free red blood cells in the skin. The term covers a wide range of skin lesions from small pin point (petechiae < 5mm) haemorrhagic spots to larger purpura (>5 mm). Purpura usually imply malfunctioning clotting mechanism (vitamin K and C deficiency in ED patients) whereas petechial lesions are due to reduced platelet count). Characteristically these do not blanch on pressure. Full blood count, erythrocyte sedimentation rate and clotting (PT/INR) will be the initial line of investigation.

Common causes like vomiting (especially after a violent episode of vomiting) to be ruled out. Liver disease, renal failure and other serious condition like thrombocytopenia, and pancytopenia are the most likely causes in severe anorexia nervosa patients. Thrombocytopenia would be confirmed by reduced platelet count, but RBC and WBC (confirmed by full blood count) will be within normal limit whereas the pancytopenia would show a reduction in all of these.

Thrombocytopenia: If platelet count is less than 20×10^9 or no haemorrhagic manifestations—no treatment required apart from observation. If the level is less than this or haemorrhages present, treat with oral steroids (gradually tapering), ± immunoglobulins and immunosuppressants. Main treatment is to improve the nutritional status of the individual.

Pancytopenia or aplastic anaemia: improving the nutritional status, treatment of infections with antibiotics and antifungal agents, oral steroids ± immunoglobulins.

Pressure Sores:

Seen in extremely low weight patients with anorexia nervosa who were immobile for a period. Other factors like low haemoglobin, low total protein and hypokalemia could contribute to the development of pressure sores. Management is by improving nutritional intake, close monitoring of the blood profile and rectifying any abnormalities. Sores are usually treated by regular wound care, improving mobility and treatment of any infection.

Itching with no skin lesion:

Routine investigations to exclude causes like chronic liver disease, renal failure, iron deficiency and hypothyroidism.

Dr. Murali Sekar, Dr. Krishnakumar Muthu

Cardiovascular and Respiratory symptoms and signs:

Chest Pain:

Chest pain should be taken seriously in eating disorder patients as cardiac arrest is one of the commonest causes of death in ED patients.

Symptoms, Investigations and Treatment in Eating Disorders

Causes	Characteristics	Investigations	Intervention
Angina	Central chest pain ± radiating to jaw and either arm (left usually), intermittent, lasting for <30 min and relieved by rest or nitrates.	No ↑ troponin after 12 hours, no ECG changes	Urgent referral to emergency department for investigations and management.
Myocardial infarction	Central chest pain ± radiating to jaw and either arm (left usually). Continuous, typically over 30 minutes not relieved by rest or nitrates	↑ ST segment in ECG.	Urgent referral to emergency department for investigations and management
Oesophagitis and Oesophageal spasm	The patient typically experiences pain when supine (lying flat), after food, alcohol etc.	No ↑ troponin after 12 hours, no ECG changes	Urgent referral to emergency department for investigations and management.

			Usually treated with PPI medications and calcium antagonists.
Oesophageal rupture	Severe acute pain immediately after a bout of vomiting.	No typical changes but can occur. Presentation will be very acute.	Urgent referral to emergency department for investigations and management.
Pulmonary embolus	Sudden breathlessness, tachycardia along with history of immobility and previous history of emboli.	Confirmed by CT pulmonary angiogram.	Urgent referral to emergency department for investigations and management. Usually treated with heparin followed by warfarin

Pneumo-thorax	Pain in the centre or side of the chest with abrupt breathlessness. ↓ breath sounds on auscultation, Palpation may reveal tracheal deviation and hyper-resonance on percussion.	Chest X ray	Urgent referral to emergency department for investigations. Treatment by release of tension through insertion of venflon in the second intercostal space, mid-clavicular line.
Dissecting aortic aneurysm	Tearing pain, radiating to back. Absent pulses, sudden alarming drop in BP.	CT or MRI Scan	Urgent referral to emergency department for investigations. Urgent surgical intervention is often required.

Chest wall pain	Superficial chest pain, tenderness on chest wall.	No specific changes in blood tests, ECG or CT.	Urgent referral to emergency department for investigations to rule out other serious conditions.
Gastro-oesophageal reflux/ gastritis	History of recurrent vomiting. Can also be seen in very low weight patients. Pain is more central and just below the rib cage, lasting for more than 30 minutes, Often worsened by food.	Normal troponin after 12 hours, no ECG changes. Oesophagitis on endoscopy.	If not clear, urgent referral to emergency department for investigations. Treated with PPI and calcium antagonisis.

Pancreatitis	Reported in very low weight patient. Risk could be elevated in alcoholics. Pain is mid-epigastric area, radiating to back with nausea and vomiting.	↑ serum amylase CT scan may reveal pancreatic pseudocyst.	Urgent referral to emergency department for investigations and treatment.

Dr. Murali Sekar, Dr. Krishnakumar Muthu

Dyspnoea (difficulty in breathing)

<u>Acute:</u> If a patient experiences sudden difficulty in breathing (occurring over few seconds) without chest pain suspect pulmonary embolism and pneumothorax. Other conditions like acute bronchitis (common chest infections), acute left ventricular failure, arrhythmia or electrolyte imbalance (especially hypokalemia) can present with sudden difficulty in breathing. Acute pulmonary oedema may present with dyspnoea, cough with pink frothy sputum (the pink coloration is due to the presence of blood mixed with oedematous fluid from lungs).

<u>Chronic:</u> Orthopnoea is defined as difficulty in breathing when lying down. Paroxysmal nocturnal dyspnoea (PND) happens when patient lies down in bed at night or due to sudden bronchospasm at night. Usually requires blood tests, ECG and chest X ray to make a definite diagnosis. Pulmonary oedema (due to cardiac failure—a number of causes could lead to this condition in ED patients) and cardiac arrhythmia are generally relevant in ED settings. Other causes like pre existing asthma and chronic obstructive pulmonary disease (COPD) should be considered.

Pulmonary oedema: usually presents with fatigue and exertional dyspnoea, whereas PND presents with palpitations, chest pain, dizziness, hypotension and bradycardia.

Syncope:

Syncope is defined as sudden loss of consciousness. This is usually due to abnormal aberrant brief electrical activity in central nervous system e.g vasovagal attack, cerebrovascular accident or sudden and temporary decrease in a cardiac output usually due to decline in blood pressure.

- Vasovagal attack—simple fainting can occur in ED patients precipitated by sudden emotional shock, pain, prolonged standing or excessive exercise.
- Postural hypotension—sudden loss of consciousness when getting up from lying or sitting position. Confirmed by a drop in blood pressure of >20 mm Hg from reclining to standing. Patients should be informed of the risk and to avoid the precipitant activity. The clinician should provide psychoeducation about prolonged standing and strenuous exercise. Support stockings and oral steroids (Fludrocortisone) could be helpful.

Hypoglycaemia—This is another significant cause of syncope in ED patients. This condition can occur in sufferers of anorexia, bulimia as well as binge eating disorder patients. Diabetes increases the risk exponentially in these individuals. Typically blood sugar would be <2 mmol/L (36 mg/dl). Hypoglycaemia is managed by intake of glucose. Glucagon may be required. Adjust the dosage of insulin and oral antidiabetic medication. Insulin-purging could result in deranged glucose metabolism leading to hypo and hyperglycaemia. Other medical conditions like aortic stenosis, CVA, pulmonary embolism should be ruled out.

Cyanosis:

Peripheral: bluish discoloration of hands but not tongue; Central : bluish discoloration of tongue but not hands. Peripheral cyanosis occurs in anorexia nervosa when the cardiac output is compromised due to arrhythmia, electrolyte abnormalities. Onset is usually minutes or hours accompanied by difficulty in breathing. Treatment requires specialist intervention with O2, diuretics and nitrates. Septicaemia can cause peripheral cyanosis too. Central cyanosis is not directly caused by eating disorders. If present suspect pre-existing cardiac conditions like Ebstein's anomaly, tetralogy of Fallot, haemoglobinopathies etc.

Hoarseness:

Hoarseness could be due to chronic laryngitis, inhaled steroids (as a treatment for asthma) or serious conditions like laryngeal carcinoma. Persistent hoarseness (few weeks to few months in duration) should be investigated.

In ED settings, hypothyroidism could be a significant differential diagnosis. This is usually seen following the cessation of externally administered thyroid preparations. Onset of hoarseness is usually slow (over months or years). Other symptoms are fatigue, puffy face, cold intolerance, bradycardia and slow deep tendon reflexes. Investigation would reveal ↓Free T4 and ↑ TSH. Laryngoscopy would reveal swollen vocal cords.

Bilateral oedema:

Causes	Relevance to ED	Other features	Treatment
Congestive cardiac failure	Can occur	Dyspnoea, loud S2, ↑ JVP, hepatomegaly. Onset over months	Specific treatment for CCF required e.g diuretics, O2 etc. Nutritional status should be improved.
Low albumin status	Commonly seen	Onset over months. Often with generalised oedema. Low serum albumin.	Improve nutrition. Diuretics are rarely used.
Refeeding syndrome	Seen during refeeding	Onset relatively quickly. History of sudden weight gain during refeeding. Associated ↓ in K, Mg and/or Po4 levels	Rectifying any abnormality. Weight gain through balanced meal at a rate of 0.5 – 1.0 Kg/week. Diuretics (Spironolactone) are rarely used.

| Bilateral cellulitis, Inferior vena caval obstruction due to prolonged immobility | Can occur but uncommon. | Other signs of inflammation, onset over few days, history of immobility etc. | Treat the cause |

Pulse:

Tachycardia (pulse rate is > 120 beats per minute): In ED setting, suspect severe anaemia, heart failure, electrolyte abnormalities (e.g hypokalemia) and thyrotoxicosis factitia. No specific intervention is required in mild cases. Severe and persisting symptom should be investigated with full blood count, U&E, CRP, D-dimer, troponin, arterial blood gas, ECG and chest X ray. Manage according to the underlying cause.

Bradycardia (pulse rate <60 beats per minute): Commonly seen in patients who exercise heavily. Care should be taken to find about the onset of bradycardia (rapid or insidious), history of exercise, pre-existing heart conditions etc. Hypothermia (core temperature <35 degrees Celsius (95 Degree Fahrenheit) is associated with bradycardia. If the temperature is < 33 degrees Celsius (91.4 degrees Fahrenheit), hypothermia will be accompanied by reduced rate of breathing—less than 10 per minute). Hypothyroidism, myocardial infarction or myocardial ischaemia, electrolyte abnormalities are some of the other relevant conditions to be considered in ED settings.

Irregular pulse: could be due to atrial fibrillation, ischemic heart disease, atrial or ventricular ectopics. Hypokalemia is often the reason for irregular pulse in eating disorder patients. Further investigated by ECG, cardiac monitoring, troponin levels etc. Treatment according to the cause.

High volume Pulse: There will be a high bounding feel to the pulse. This is confirmed by the large difference between systolic and diastolic blood pressure (i.e pulse pressure > 40 mm Hg). Severe anaemia (folate and/or B12 deficiency) is a common cause. High bounding pulse can also be seen in bradycardia when the heart is functioning normally to compensate with increased pressure of pumping.

Low volume pulse: Pulse pressure (difference between systolic and diastolic pressure) is <20 mmHg. Dehydration, inadequate cardiac contractility (in extremely low weight patients) to be looked out for in ED patients.

Blood Pressure:

Hypertension: definition varies but commonly accepted as systolic value of > 150 mm Hg and diastolic value of > 90 mm Hg. No direct association with eating disorders but mindful of essential hypertension, secondary hypertension following renal disease or long term intake of oral contraceptive pills prescribed to treat osteoporosis.

Hypotension: There is no universal definition but systolic blood pressure is < 90 mm Hg and diastolic blood pressure is < 60 mm Hg. Low volume pulse often presents with hypotension.

Postural hypotension: fall in systolic blood pressure by more than 20 mm Hg and the pressure fails to come up within the next minute. Usually accompanied by dizziness and syncope. Causes are similar to hypotension in ED settings, but other medical conditions (e.g autonomic neuropathy, multiple sclerosis) can cause postural hypotension.

Haemoptysis (sputum streaked with blood): Care should be taken to differentiate haemoptysis from haematemesis. Haemoptysis is blood tinged with sputum or other lung secretions and does not appear as frank blood. Haematemesis can present as frank blood as well as with food or other stomach secretions. The volume also helps to distinguish. Haemoptysis is usually small in volume (often streaks of blood) compared to haematemesis in which case the blood loss could vary from small to copious. Symptoms like hypotension, palpitations are usually associated with haematemesis.

This condition is not directly linked to eating disorders. Increased susceptibility to infection in anorexia nervosa could lead to bronchitis which is a common cause for haemoptysis. Other conditions like upper respiratory tract infections, pneumonia, bronchiectasis, lung abscess can cause haemoptysis.

Blood Gas changes:

Often patients who are low weight (BMI <14) complain of cold extremities. Other signs like cyanosis, tremors and muscle twitching need to be checked. Pulse oximetry would be particularly valuable in such cases, and following findings could be observed:

Hypoxic: Mild when oxygen saturation is between 85%-92%

Severe when oxygen saturation is less than 85%.

Blood gas analysis will reveal a PaO2 of < 8 kPa. Severe hypoxia is often associated with restlessness, confusion or unconsciousness. Treatment is usually with controlled oxygen administration and treatment of the precipitating cause.

Hypocapnia: In this condition carbon dioxide level is low due to various causes. Common cause in psychiatric practice is hyperventilation as a result of anxiety. Usually require blood gas analysis. PaCo2 < 4.5 kPa along with symptoms of dizziness, paraesthesia and tachypnoea (increased rate of breathing) would confirm the diagnosis. Treatment is re breathing into a paper bag. Sometimes supplemental oxygen may be required. Addressing the anxiety is crucial to prevent relapse.

CO2 retention: could be seen in very thin anorexic patients who do not have good respiratory muscle mass to carry out the expiration and inspiration. Signs like warm hands, high volume pulse, muscle twitching, headache, confusion should raise suspicion in these individuals. PaCO2 is usually > 6.5 kPa. Symptoms tend to occur when PaCO2 is >8.0 kPa. Treatment involve O2 delivery and improving the nutritional status.

Other conditions like hypokalemia, hyperkalemia, hypocalcemia and morbid obesity can cause CO2 retention.

Bradypnoea

Respiratory rate needs to be counted for one full minute and use distraction techniques (e.g giving an impression of counting pulse) since direct observation could affect the rate and rhythm of respiration. If the respiratory rate is < 10/min. the patient would be generally semi/unconscious requiring CPR and other urgent medical interventions.

Very high CO2 (due to inefficient respiration secondary to poor respiratory muscle mass), hypothermia (exposure to cold in combination with immobility in morbidly low weight AN patients) are the causes of bradypnoea. Treatment would require O2 (to keep the oxygen saturation between 88-92%), gentle rewarming (using space blankets, IV fluids, bladder irrigation etc) depending upon the cause. Suspect other causes like drugs e.g opiates, alcohol, benzodiazepines, head injury.

Kyphosis and Scoliosis

Kyphosis: spine curved forward

Scoliosis: Spine curved laterally (on either side)

Scoliosis is observed in ED patients with osteoporosis and collapse of vertebrae. Anterior collapse of vertebrae will result in kyphosis and collapse on either sides of the vertebrae could cause scoliosis. Usually requires CXR and CT/MRI to confirm the diagnosis.

Management: Physical examination to detect any reduction in the chest wall movements. Blood gas analysis to examine respiratory functions. The tests need to be done in the night time when the patient is sleeping, since the decubitus posture or lowering of muscle tone during sleep could affect the overall respiratory function. Non invasive ventilatory support is indicated if PaO2 is < 8 kPa with PaCO2 > 6.5 kPa.

Similar clinical picture is also seen in obesity hypoventilation syndrome. Here, the respiratory excursion of the chest wall is impaired due to heaviness. This condition requires similar management along with weight reduction.

Neurology—Signs and Symptoms:

Confusion

This is characterised by altered level of consciousness, failure to register and recall memory to a varying degree, disorientation to time, place and person. They may show alteration in the motor activities e.g restless or withdrawn and drowsy. A number of conditions in ED setting could cause this presentation.

Thiamine Deficiency: look for other symptoms—nystagmus, ataxia and ophthalmoplegia.

Hypothyroidism: cold intolerance, lethargy, hoarseness in the voice, history of thyroid supplement use (current use would suggest under-correction, whereas the past use indicates altered thyroid function).

Hypoglycaemia: irritability, difficulty in concentrating, sweating, lethargy, tachycardia etc.

Infection: look for other signs like fever, rigors, cough and other symptoms to reveal the source of the infection e.g dysuria may suggest urinary tract infection.

Dizziness

One of the common symptoms in AN patients. Postural hypotension, anaemia, hypoxia, hypoglycaemia and various electrolyte and vitamin deficiencies could cause this non-specific symptom. Aim is to determine the cause and assessing the seriousness. This would inform the necessity for urgent intervention and also the nature of the intervention. For example, in hypoglycaemia with a low blood glucose chance for the individual to develop fits and lose consciousness is high. Hence urgent intervention to improve the blood glucose is essential. Low weight patients might not experience hypoglycaemic symptoms so periodic blood glucose measurement is essential. Psychoeducation about the apparent lack of symptoms will be required in ED setting.

Other conditions like anaemia take time to improve. Patients should be informed of this. Refer to relevant sections in the book to read about the management of these conditions.

Dizziness should be differentiated from vertigo which means a sensation of movement of oneself and/or the surroundings. Vertigo usually involves middle ear, but other structures like mid-brain may also be the source.

Convulsions or Fits

Gather history about aura, loss of consciousness, tonic-clonic movements of the limbs, confusion following the episode, nutritional intake before the episode, any previous attacks, any known precipitating factors e.g period of starvation, fluid restriction and recent electrolyte changes. Also, inquire about medication history as some medication could increase the potential or discontinuation of some medications can increase the risk of fits.

The following is some of the common causes in ED settings:

Hypoglycaemia: Look out for other symptoms, relation to food intake and any corresponding blood glucose estimation. Usually blood glucose would be < 2 mmol/L (36 mg/dl). Initial management with oral glucose gel if he/she can take it orally. Alternatively, try 50% glucose IV. Glucagon 1 mg IV or IM is commonly used but in severely underweight patient since the glucose store may be seriously depleted already, this may not be as helpful as Glucose administration. In case of hypoglycaemia during re-feeding, the glucagon could be beneficial.

Hypotension: Can occur following a period of acute starvation or fluid restriction. Associated low blood pressure and tachycardia would be supportive of the diagnosis. Still, other conditions need to be excluded. ECG is also essential. In severe cases, the fit might have been brought on by sudden transient loss of blood supply to brain following brief cardiac arrest. Close monitoring in intensive care setting, IV fluids, central venous line, cardiac support may all be required.

Electrolyte Imbalance: Severe lowering or significant increase of electrolytes like sodium, magnesium or calcium can cause convulsion. Please refer to relevant sections in the book for information about associated symptoms, blood levels and their management.

Be mindful of 'functional fit' or 'pseudo-fit'. History and lack of corresponding signs and/or abnormal blood tests would suggest this possibility.

Transient neurological deficits:

Symptoms and signs e.g speech difficulty, weakness of facial muscles or limb muscles with or without altered sensorium lasting for < 24 hours. In ED setting, transient hypotension, hypoglycaemic episodes are the common causes. In addition, hyponatraemia should be suspected as severe hyponatraemia (< 120 mmol/L) can present as transient neurological deficits. This could be a warning sign and more serious symptoms like convulsions and coma may follow soon.

Tremors

Fine tremors: may be made more noticeable by placing a sheet of paper on the fingers of an out-stretched arm. Anxiety, hypoglycaemia, alcohol withdrawal, thyrotoxicosis (factitia) are the common causes in ED settings. Benign essential tremor (onset is usually late in life) is another form of fine tremors. This condition is not pathological.

Coarse tremors: These are not directly linked to ED conditions. These are more alarming than fine tremors. Causes include liver failure and carbon dioxide retention which can occur in later stages of severe malnutrition.

Squat test:

A commonly used test to assess the severity of malnutrition. The test involves assessing the power of the group of muscles around hips and knees. When arms are used to support getting up, this group of muscles are also tested.

Poor musculature as a result of malnutrition is the reason for the failure. Some rare conditions like polymyositis and various forms of neuromyopathy and diabetic amyotrophy can cause failure of squat test. Rarely thyrotoxicosis and osteomalacia can cause failure of squat test. If a patient fails squat test and appears not to be thin or weak enough, then suspect these other conditions. In the other conditions, muscle mass is preserved in the early stages.

Gastroenterology—signs and symptoms

Swollen salivary glands

Parotid and submandibular glands are often involved in purging behaviour. Hypertrophy of the salivary glands occurs due to hyperactivity of salivary glands (due to repeated vomiting). Repeated vomiting also leads to increased pressure in these glands due to repetitive and later, sustained increase in pressure in papillae of the glands. Patient may complaint of swelling in the cheeks but at times present with inflammation of these glands.

Treatment usually involves cessation of purging behaviour, warmth over the swollen glands, mouth irrigation (saline or lemon mouthwash) in non inflammatory stages. If infected treatment with antibiotics along with other measures would be necessary.

Dental Caries

Dental caries can result from repeated erosion of dental enamel by acid content of the vomitus. Membranes of the oral cavity, teeth enamel are not designed to withstand the acid erosion. Hence the clinical manifestation such as dental caries, poor oral hygiene, opportunistic infections (normal bacterial flora of the mouth is altered by the acid in the vomit) are the oral complications of repeated vomiting.

Referral to a dentist for appropriate intervention is often necessary. Psycho-education needs to be offered to those who indulge in recurrent vomiting about the dangers of their behaviour.

Simple advice like gentle brushing of teeth after every episode of vomiting would be helpful. Excessive use of brushing or stronger mouthwash preparations could damage the already weak enamel.

Vomiting

It will be uncommon for ED patients who make themselves sick to present solely with this symptom. Often the clinician needs to elicit the history tactfully. They might be frugal with the information i.e number of times, how long after the meal, using fingers etc. Sometimes patients 'chew and spit' food. Understandably these symptoms could be very embarrassing thus would not be volunteered. Tactful elicitation would lead to offer of suitable psycho-education and other interventions.

ED professionals should be aware of other causes of vomiting, which can occur in ED patients. As a standard investigation, self induction must be ruled out. Following are some of the medical/surgical reasons for vomiting

<u>Vomiting associated with difficulty in swallowing food:</u>

Oesophageal stricture: suggested by the presence of undigested solid food in vomit even after a considerable period lapsed since they have eaten.

Oesophageal carcinoma: worsening degree of difficulty in swallowing i.e starting with solids progressing to liquids. Other symptoms like weight loss will be contrary to patient's intention (to gain weight). Pain, bleeding are late complications, and they usually suggest terminal stage of the cancer.

Gastric carcinoma: increased satiety after a small meal (which is often seen in ED patients too), bloating, fullness not confined to dietary intake etc. Pain and bleeding are usually late complications.

Tumours of small intestines: Conditions like lymphoma could lead to intestinal obstruction (partial/full) leading to vomiting and loss of weight. If the vomit contains bile, the obstruction is distal to the opening of biliary duct (in the second part of duodenum).

Other less serious conditions like inflammation of Oesophagus (can arise due to recurrent vomiting especially at the junction between Oesophagus and stomach) can cause vomiting with difficulty in swallowing. Symptoms like heartburn, regurgitation would raise the suspicion.

Vomiting with abdominal pain

Conditions like acute appendicitis, gastroenteritis, acute pancreatitis, hepatitis, renal colic are the causes of vomiting with abdominal pain. Pain will be very noticeable in these conditions. Hence the objective evidence for pain could distinguish ED patients' self induced vomiting and vomiting due to the other conditions. These conditions would require urgent medical/surgical intervention.

Jaundice

Jaundice is characterised by yellow discolouration of conjunctivae and skin. A bilirubin level of > 35 micromol/L (2 mg/dl) will lead to this yellowish discoloration. In ED practice, patients who consume excessive yellow vegetables and fruits might manifest carotinaemia. A normal serum bilirubin level along with dietary history would establish the right diagnosis.

A number of conditions with varying degrees of morbidity and mortality is linked with this clinical manifestation. Starting point would be liver function tests. Based on the findings a referral to the specialists can be made for further treatment.

Haematemesis

Haematemesis means the presence of blood in the vomiting. This also indicates that the source of bleeding is from the upper intestinal tract i.e oesophagus, stomach, first and the second part of the duodenum).

In ED setting history of repeated vomiting, any recent increase of such behaviours along with haematemesis would indicate serious complications like oesophageal or gastric rupture (Mallory-Weiss tear). Other conditions like oesophageal varices, peptic ulcer, gastritis, oesophageal or gastric tumours should be excluded. Further investigations to rule out bleeding disorders (low platelet count, vit.K deficiency) should be carried out.

Usually large quantity of fresh blood would suggest active bleeding needing urgent surgical intervention. Mallory—Weiss tear often present as a streak of blood in the vomit indicating minor oesophageal trauma. Such warning sign should be taken seriously, and the patient should be informed of the imminent risks. Refer the patient to gastroenterologist to assess the damage and need for further intervention.

Melaena (black stool) and coffee ground blood in the vomit is due to alteration of the fresh blood by intestinal secretions or gastric secretions respectively. Iron supplementation can also cause the colour of the stool to change to black resembling melaena.

Distended abdomen/bloating:

Often complained by patients especially anorexia nervosa sufferers who tries to improve their nutrition. This symptom needs further evaluation to learn whether this is an actual physical ailment or an aspect of body image distortion or a combination of both. Addressing both elements are crucial for successful management of this condition. Psycho-education about the failure of movement of food due to poor gastric/intestinal musculature would explain the nature of the symptom to the patient. Importance of improving protein intake through a balanced meal to strengthen the muscle layers should be offered. A strong smooth muscle layer in the stomach and intestine would enable the food movement.

Other cause like flatulence, reduced or absent bowel movement following cessation of laxatives would also need to be considered as causes for this symptom.

Miscellaneous

Secondary amenorrhoea

- Primary amenorrhoea: when a patient never had menstruation. In ED setting, this can be seen when AN started in very young age i.e before menarche.
- Secondary amenorrhoea: absence of menstruation for a period of three consecutive months in a person who had periods prior to the cessation.
- Anorexia nervosa with a BMI of < 17.5 is the commonest cause in ED setting. The following is some of the other causes:
- Pregnancy: Is possible even when a person is not menstruating.
- Menopause: especially if the patient is in the appropriate age group (>40 years).
- Premature Ovarian failure: This condition is observed at times, following a prolonged period of amenorrhoea. Period fails to appear even after the weight restoration. Blood test will reveal ↑ LH, ↑ FSH, ↓ Oestradiol. Atrophic changes will be seen in ovaries on ultrasound.
- Thyrotoxicosis can present with amenorrhoea.

Minerals
and Trace elements

Introduction

Until recently (2005), the minerals are considered to be those substances formed only be geological processes. The substances like urinary calculi and oxalates were excluded from the list of minerals since it involved a biogenic process. Now these products are also considered as minerals. Dietary minerals are the chemical compounds required by living organisms other than the four major compounds carbon, oxygen, nitrogen and hydrogen. Most of these dietary minerals enter the body, not as simple chemical compounds but large aggregates requiring breakdown by digestive processes and the action of intestinal flora. Some of this breakdown also occurs outside the body and enters the human body as part of the ecological food chain.

Diet can meet all the body's chemical element requirements. Supplements are used when some requirements (e.g., calcium, which is found, mainly in dairy products) are not adequately met by the diet. Chronic or acute deficiencies also arise from pathology or injury. Research has supported that altering inorganic mineral compounds (e.g carbonates, oxides) by reacting them with organic ligands (e.g amino acids, organic acids) improves the bioavailability of the supplemented mineral.

Certain minerals and trace elements are considered essential'. For a mineral or a trace element to be considered essential for the human body the following characteristics should be fulfilled:

- It should be present in healthy tissues.
- The concentration of the chemical should be relatively constant amongst the individuals of the species.

- Deficiency should result in biochemical abnormalities and these abnormalities should be equal amongst individual members.
- Replacement of the deficit should lead to observable correction of the biochemical changes.

In order to maintain the concentration, most of these minerals need to be taken in sufficient quantities. For minerals, this would be in the order of grams and milligrams. For trace elements, this will be micrograms.

In human biochemistry, a trace element is a dietary mineral that is needed in minute quantities for the proper growth, development, and physiology of the organism. The difference, for all the practical purposes in eating disorders between trace elements and minerals, is in their recommended daily intake. Human beings depend on at least nine trace elements--iron, zinc, copper, manganese, iodine, chromium, selenium, molybdenum and cobalt--for optimal metabolic function. These elements serve a variety of functions including catalytic, structural and regulatory activities in which they interact with macromolecules such as enzymes, pro-hormones, and biological membranes.

Apart from the minerals and trace elements, there are other chemical substances that are considered to be of biological importance. They do not satisfy the characteristics given in the previous page. This include nickel, boron, vanadium and cobalt.

Human body mass roughly consists of:

Oxygen (65%), Carbon (18%), Hydrogen (10%), Nitrogen (3%), Calcium (1.5%), Phosphorus (1.0%), Potassium (0.35%), Sulphur (0.25%), Sodium (0.15%), Magnesium

(0.05%), Iron (70%), Copper, Zinc, Selenium, Manganese, Cobalt, Molybdenum, Fluorine, Chlorine, Iodine, Lithium, Strontium, Aluminum, Silicon, Lead, Vanadium, Arsenic, Bromine (trace amounts)* Starvation activates mechanisms in the body to conserve the use of minerals and trace elements. Self-prescribed nutrient supplements can result in laboratory values that appear within normal limits. Care to be taken to avoid these pitfalls. Also, the deficiency of minerals and trace elements should not be seen as acute problem but as a long standing one which needs long term assessment and treatment and regular monitoring by various professionals including dieticians. Blood test alone may not be sufficient.

* H. A. Harper, V. W. Rodwell, P. A. Mayes, Review of Physiological Chemistry, 16th ed., Lange Medical Publications, Los Altos, California 1977.

Sodium

Role of Sodium

- Sodium is an electrolyte that plays an important role in water maintenance in the body
- Maintains blood pressure
- Important in muscle functioning and conduction of nerve impulses

Rich in:

Sodium is found in table salt, baking soda, monosodium glutamate (MSG), various seasonings, additives, condiments. Meat, fish, poultry, dairy foods, eggs, smoked meats, olives, and pickled foods are rich sources of sodium.

ED specific points of interest

Sodium intake is recommended to be less than 3,000 milligrams daily. One teaspoon of table salt contains about 2,000 milligrams of sodium. Serum sodium is usually normal in patients with an eating disorder. Increase in sodium level stimulates thirst, and a reduction in level inhibit thirst centre.

Normal Range

135-145 mmol/L [135-145mg/dl]

Hyponatremia
(Sodium <135 mmol/L [135 mg/dl])

ED specific points of interest

Most patients with low sodium could be asymptomatic. Symptoms usually begin to appear when sodium level goes to less than 120 mmol/L [120 mg/dl]. At this stage, it is more difficult to treat. Hence beware of low sodium even when patient has no symptoms.

Symptoms are non-specific like headaches, lethargy, nausea. In severe cases patients may present with non-specific neurological e.g dizziness, loss of sensation and or gastrointestinal symptoms e.g abdominal cramps etc. Seizures, coma and death can occur.

Rate of decline of sodium level can determine the onset of symptoms i.e faster the rate of drop earlier and more likely are the symptoms

Three types:

- Low sodium with a low volume of fluid (ECF) e.g Renal causes like abuse of diuretics and extra renal causes like vomiting and diarrhoea
- Low sodium with normal fluid level e.g psychogenic polydipsia (the excessive drinking of fluids like water and soft drinks)
- Low sodium with increased fluid volume e.g when concurrent low protein, liver or heart failure exists in eating disorder patients.

The presence of following indicates poor prognosis: presence of symptoms as opposed to absence of symptoms, concurrent infections, signs of respiratory failure

Three deciding factors for the treatment of Hyponatremia:

- presence of symptoms,
- acute or chronic (onset less or more than 48 hours),
- presence of low blood pressure (BP Systolic <90 and Diastolic <60)

Medical conditions like congestive cardiac failure, liver or renal impairment (as evidenced by blood tests), hypothyroidism and Addison's disease need to be investigated and ruled out or treated accordingly. A Sodium level of <130 mmol/L [30 mg/dl] is seen in chronic vomiting or laxative abuse. Hence urine sodium level could be used to investigate the presence or absence of these symptoms. Medications like SSRIs and certain types of diuretics can lower the sodium level.

Management of Hyponatremia with Hypovolaemia

If symptomatic, rule out other causes ↓	>135 mmol/L [>135 mg/dl] ↓	Treat other causes as required ↓		
Lethargy, headache, nausea, thirst, muscle cramps, postural dizziness ↓	120–135 mmol/L [120-135 mg/dl] ↓	Oral electrolyte and glucose mixtures. Increase the salt intake (60—80 mmol/day) ↓		COMMUNITY
Confusion, myoclonic jerks, generalised convulsions and coma	<120 mmol/L [<120 mg/dl]	IV Fluids with Potassium supplements 1.5—2.0 L of 5% Dextrose with 20 mmol of Potassium. Add 1 L of 0.9 % Saline over 24 hours for any measurable loss	MEDICAL WARD	EDU

Hypernatraemia

ED specific points of interest

- Much rarer in eating disorders, but one study reported hypernatremia to be common in children with anorexia nervosa before the commencement of treatment i.e baseline evaluation*.
- Mostly due to water deficit especially when patients restrict water intake.
- More likely to occur in hot weather due to excessive loss of water.
- Some antibiotics e.g piperacillin contains a high amount of sodium.
- Attempt to self correct low sodium could result in hypernatremia.
- Hypernatremia does not always indicate an increase of total body sodium. Reduction in body water could lead to an apparent increase in sodium level in the serum.
- Symptoms are non-specific
- Nausea, vomiting, fever and confusion may occur. In severe hypernatremia (>170mmol/L [>170 mg/dl]) convulsions can occur.
- Medical conditions like diabetes insipidus, iatrogenic causes (treatment induced) should be ruled out.
- If serum level is raised, concurrent urine and plasma osmolality and sodium level should be measured to find out the cause for hypernatremia.

* Evaluation of parameters in with (I). Nogal P., Pniewska-Siark B., Lewinski A. EmbaseNeuroendocrinology Letters. 29 (4) (pp 421-427), 2008.

Management of Hypernatraemia

Symptoms	Sodium level	Treatment	Setting
Mostly non-specific like nausea, vomiting. Fever can be a presenting feature.	145-150 Mmol/L	Investigate with simultaneous urine and plasma osmolality and sodium measurements. Improve water intake. ↓	COMMUNITY
↓ symptoms are polyuria, polydipsia and thirst may become more prominent (if there is underlying diabetes insipidus) ↓	↓ 150-170 mmol/L ↓	IV Fluids 5% Dextrose or 0.45% Saline (large volume may need to be given) ↓ IV Fluids 0.9% saline Aim is to bring down the level to normal range in about 48 hours.	EDU
Confusion and coma	>170 mmol/L	Avoid too rapid correction as this leads to cerebral oedema	MEDICAL WARD

Potassium

Role of Potassium

- Important for conduction of electrochemical potentials along with sodium, chloride, calcium, and magnesium.
- Crucial to heart function and plays a key role in skeletal and smooth muscle contraction

Rich in:

All meat products, some types of fish (e.g) salmon, cod and many fruits, vegetables, and legumes. Dairy products are also good sources of potassium.

ED specific points of interest:

- Daily intake varies between 80-150 mmol/L [80-150 mg/dl]
- Serum potassium level is determined by K uptake into cells, excretion by kidneys and losses e.g vomiting, diarrhoea
- Uneven distribution of potassium means a shift of 1% could result in 50% variation in the plasma concentration.
- Vomit contains 5-10 mmol/L [5-10 mg/dl] of Potassium; Diarrhoea stool contains 10-30 mmol/L [10-30 mg/dl] of Potassium
- Total potassium in the body is between 3000-3500 mmol/L [3000-3500 mg/dl] • Potassium level is affected by hormones e.g insulin, aldosterone, drugs e.g theophylline, diuretics; conditions e.g acidosis (K↑) and alkalosis (K↓)

Normal Range: 3.5-5.5 mmol/L [3.5-5.5 mg/dl]

Hypokalemia

ED specific points of interest

- Common in eating disorder conditions especially with purging behaviour
- Increased renal excretion due to abuse of diuretics (thiazide and loop diuretics) can lead to hypokalemia. In such cases urinary K> 20 mmol/day [>20 mg/dl].
- Purging behaviour is usually associated with urinary K < 20 mmol/day [>20 mg/dl].
- Concurrent heart failure and liver failure increase the risk of hypokalemia.
- Body usually adapts to low potassium intake by reducing aldosterone secretion but this is usually ineffective.
- Mechanism of development of hypokalemia in vomiting and diarrhoea:

Vomiting & Diarrhoea
↓
Loss of fluids, Na ions
↓
Increased Aldosterone and Angiotensin II levels
↓
Secretion of K and H ions in exchange for Na ions in the collecting duct of nephrons
↓
Hypokalemia

- Levels of Na, K and H are closely linked. K ions and/or H ions are exchanged for Na ions. But K and H ions are interchangeable depending upon the relative abundance/scarcity of one ion over the other. Body tends to retain the ion which is lesser than the other i.e In acidosis, H ions are more hence kidneys will preserve K and excrete H ions in exchange of Na; In alkalosis where H ions are relatively scarce the reverse happens.

- Common symptoms are palpitations, muscle weakness (Negative SUSS test). More serious symptoms like syncope, chest pain can occur.

- In case of treatment resistance suspect hypomagnesemia.

- Ingestion of potassium can cause insulin release resulting in lowering of blood sugar level.

Management of Hypokalemia

Symptoms	Serum K Level	Treatment
Palpitations, muscle weakness Investigate. Serum K level is reflective of body store ↓ Measure Urinary K level, Mg level ↓ ECG (QTc prolongation) and other close monitoring	>3.5 mmol/L [>3.5 mg/dl] ↓ 3.0—3.5 mmol/L [3.0-3.5 mg/dl] ↓ 2.5—3.0 mmol/L [2.5 -3.0 mg/dl] ↓ 2.0—2.5 mmol/L [2.0-2.5 mg/dl] ↓ <2.0 mmol/L [<2.0 mg/dl]	? K supplement 20-40 mmol/day [20-40 mg/dl] ↓ Oral K supplement 20-60 mmol/day [20-60 mg/dl] ↓ Oral K Supplement 40-60 mmol/day [40-60 mg/dl] & consider potassium infusion ↓ K infusion 40 mmol/l [40 mg/dl] of Kcl in 1 Litre of 0.9% saline at a rate of 150 ml/hr.

COMMUNITY

EDU

MEDICAL WARD

Dr. Murali Sekar, Dr. Krishnakumar Muthu

Hyperkalemia

ED specific points of interest:

- A level of >6.0 mmol [6.0 mg/dl] usually requires treatment.
- Common drug to cause elevated level of potassium is NSAIDs. It is a common feature of chronic renal insufficiency due to various causes associated with eating disorders. Hyperkalemia in such condtions are usually ascribed to impaired K+ homeostasis. Various experimental observations suggest that the increase in extracellular [K+] actually functions in a homeostatic fashion, directly stimulating renal K+ excretion through an effect that is independent of, and additive to, aldosterone*.
- If K is high and Na is low, suspect adrenal insufficiency.
- First step in the treatment of hyperkalemia is to rule out false blood test. Usual causes are red blood cell lysis and clotted blood. In case of doubt, check both plasma as well as serum level.
- ECG usually determines the need for treatment along with rapid rise of potassium level in the blood.
- Improvement in ECG picture happens usually within 2-3 minutes of Calcium administration.
- Sodium bicarbonate (NaHCo3) is no longer recommended in the treatment of hyperkalemia.
- Muscle weakness is the usual presenting symptom. Reduced bowel movement leading to ileus paralyticus could happen.
- ECG changes: Tall 'T' waves usually precede arrhythmic changes.

* Hyperkalemia—An adaptive response in chronic renal insufficiency: Gennari F.J, Segal A.S; Kidney International. 62(1):1-9, 2002 Jul.

Management of Hyperkalemia

Muscle weakness, ileus– suspect hyperkalemia. Measure serum level ↓ ECG– Look for 'T' wave changes. If +, start treatment. ↓ Close ECG monitoring is required in ICU Setting– Arrhythmic changes could be seen.	,5.5 mmol/L [<5.5 mg/dl] ↓ 5.5– 6.0 mmol/L [5.5-6.0 mg/dl] ↓ >6.0 mmol/L [>6.0 mg/dl]	No treatment is usually required ↓ No treatment required. To commence close monitoring. ↓ IV calcium gluconate (slow infusion over 20-30 minutes in 10-20 ml of 10%IV fluid ↓ Insulin—10 U in 50 ml of 50 % glucose ± Inhaled beta-2 agonist (nebulised Albuterol 10-20 mg)

COMMUNITY

EDU

MEDICAL WARD

Magnesium

Role of Magnesium

- Magnesium is the fourth most abundant mineral in the body
- Helps to maintain normal muscle and nerve functioning, heart rhythm, functioning of the immune system, blood glucose and blood pressure within the healthy range.
- Also plays a role in protein synthesis and energy metabolism

Rich in

Magnesium is rich in legumes, nuts, whole grains, and vegetables especially spinach. Refined grains, as opposed to whole grain are deficient in magnesium.

ED specific points of interest

- About one third to one half of dietary magnesium is absorbed.
- Daily intake is around 400 mg in males and 300-400 mg in females. Girls aged between 14-18 would require significantly more magnesium intake than any other age group.
- African American diets contains less magnesium than the typical Caucasian diets.
- Most food labels do not include information about the magnesium content.
- Role of magnesium in the treatment of diabetes, hypertension, osteoporosis and heart disease gathers evidence.

Normal Range 0.7-1.0 mmol/L [1.7-2.4 mg/dl]

Magnesium Load Test:

- This test used when patients show magnesium deficiency, but their serum magnesium level is normal.
- Absorption, loss and storage of magnesium within the body are tested by this investigation.
- Steps: Patient is given a measured quantity of magnesium orally or parenterally. Serial measurements of serum magnesium and urinary excretion levels are made. The difference between the amount administered and excreted will help to confirm the diagnosis.
- If the patient retains less than 30-40% of administered magnesium, it is less likely that he/she has magnesium deficiency.
- Normally, magnesium is stored in the bones. If the patient's bone contains less magnesium, retention of magnesium will be high. This results in reduced excretion of magnesium through kidneys.

Dr. Murali Sekar, Dr. Krishnakumar Muthu

Hypomagnesemia

ED specific points of interest

- Vomiting, diarrhoea affects the magnesium levels as well as the status of gastrointestinal and renal system in the body.
- Severe magnesium deficiency can result in hypocalcemia.
- Severe deficiency is often associated with hypokalemia.
- High blood sugar in diabetes could result in increased excretion of magnesium due to increased frequency of micturition.
- Diuretics abuse cause low levels especially when the patient is on certain antibiotics e.g gentamycin, amphotericin.
- Early signs and symptoms of Magnesium deficiency:
- loss of appetite, nausea, vomiting, fatigue, and weakness
- As magnesium deficiency worsens numbness, tingling, severe muscle cramps, seizures and arrhythmia develop.
- Fatigue in the focus of eyes and short term memory loss are unique but not exclusive features of Mg deficiency.
- When the magnesium level is too low (<0.5 [1.2 mg/dl]), increasing dietary intake of magnesium might not be helpful.
- Magnesium oxide and magnesium carbonate have high elemental magnesium and bioavailability compared to magnesium sulphate and chloride preparations.

Management of Hypomagnesemia

Suspect magnesium deficiency if there are symptoms like muscle cramps, fatigue, weakness, blurring due to fatigue of eye muscles. Measure serum magnesium. Magnesium load test as appropriate. ↓ Worsening numbness, tingling sensations, palpitations, altered sensorium. ↓ Seizures, arrhythmias in ECG	0.7—1.0 mmol/L [1.7-2.4 mg/dl] ↓ 0.5-0.7 mmol/L [1.2-1.7 mg/dl] ↓ <0.5 mmol/L [<1.2 mg/dl]	Look for other electrolyte abnormalities with similar symptoms and signs. ↓ Oral Mg supplements e.g Magnesium glucoheptonate 5 ml, Magnesium gluconate 1 tablet or Ca + Mg tablets three times daily ↓ Intramuscular Mg 2 ml hourly for 6 hours followed by 2 ml every 4 hours until a sustained Mg level is observed. ↓ Intravenous MgSo4 20 mmol in 250 ml of 0.9% of saline over 4 hours and repeat as necessary	COMMUNITY MEDICAL WARD

Hypermagnesaemia

ED specific points of interest

- Hypermagnesaemia can, in theory, occur in abuse of magnesium containing laxatives.
- Care should be taken whilst correcting hypermagnesaemia since the treatment could easily lead to a level below the normal range.
- Hypermagnesaemia is extremely rare because the kidney is highly effective in excreting excess magnesium. Hence in case of renal failure, the level could go up.
- Hypermagnesaemia usually does not present solely. Commonly observed with hypercalcemia.
- Hypothyroid can rarely cause hypermagnesaemia.

Management of Hypermagnesemia

Symptoms	Serum Level	Management		
Suspect hypermagnesaemia if there are symptoms like weakness, Measure serum magnesium. ↓	1.0—1.5 mmol/L [2.4 -3.6 mg/dl] ↓	Stop administering Magnesium containing preparation ↓	EDU	COMMUNITY
Reduced tendon reflexes. ↓	1.5—2.0 mmol/L [3.6 – 4.8 mg/dl] ↓	Intravenous infusion with 10 ml of Calcium Gluconate + Dextrose solution + Insulin ± Dialysis (when level is >8 mmol/L [20 mg/dl])	MEDICAL WARD	
Respiratory paralysis and cardiac conduction defects leading to arrhythmias	>2.0 mmol/L [> 4.8 mg/dl]			

Calcium

Role of Calcium

- Strengthening bones and teeth
- Regulating muscle contraction and relaxation
- Regulates heart functioning especially conduction of electric impulses
- Essential in blood clotting
- Transmission of messages in the nervous system
- Vital role in enzyme function

Rich in

Dairy foods and calcium fortified products such as soya milk and breakfast cereals are rich in calcium. Leafy green vegetables (broccoli, spinach), fish (salmon), nuts and seeds are also rich in calcium.

ED specific points of interest

- Daily intake should be between 1000-1500 mg. Intake of more than 2000 mg is undesirable, and a possible relationship with heart disease is reported if intake is more than 2500 mg consistently.
- 200 ml of milk contains 400 mg of calcium.
- Calcium contributes to about 2% of adult weight.
- Adequate Calcium intake during adolescence would improve the adult bone calcium content through storage. Onset of eating disorders during adolescence put these patients at risk of osteoporosis straight from the beginning.
- Calcium has an extremely complex regulatory mechanism involving a number of hormones.

Hence assessment and treatment often requires interpretation of these parameters.
- Treatment of low sodium would require increased need for calcium intake.

Normal Range: 2.23-2.63 mmol/L [8.92-10.52 mg/dl]

Calcium Regulation

- Calcium is mainly stored in the bones (approximately 99%)
- Of the rest (1%), 50% is in the free (active) ionized form (1-1.15 mmol/L [4-4.6 mg/dl]), 40% is bound to protein (predominantly albumin), and 10% is complexed with anions (eg, citrate). This is one of the important reasons to measure serum albumin level. Serum albumin level can be low in patients with anorexia nervosa.
- Parathyroid hormone (secreted by the parathyroid gland, an endocrine gland) regulates the release of Ca^{2+} from bone, reabsorption in the kidney back into circulation, and increase in the activation of vitamin D3 to Calcitriol.
- Vitamin D plays a key role in the absorption of calcium.
- Calcitonin secreted from the parafollicular cells of the thyroid gland controls the release of calcium from the bone (hence tried as treatment of osteoporosis in anorexia nervosa).

Hypocalcaemia

ED specific points of interest

- Low serum calcium may be due to low magnesium or low serum albumin. Hence check these levels if low calcium is observed in the blood test.
- Serum free calcium is an additional test to confirm serum calcium.
- 99% of calcium is bound in the bones. Half of the rest circulates free (free calcium) and the rest bound with proteins in the blood.
- Acute hypocalcemia can lead to syncope, congestive cardiac failure and angina.
- Symptoms of chronic hypocalcemia include cataracts, dry skin, coarse hair, brittle nails, psoriasis, chronic pruritus, and poor dentition.
- Confirm ionised hypocalcemia, Parathormone level and Vit. D level to rule out the other important causes which could occur in any one including sufferers of eating disorders.
- Calcium citrate, calcium carbonate and calcium gluconate are preferred oral supplement preparation. Calcium chloride is used in IV preparations.

Management of hypocalcaemia

Symptoms	Severity	Treatment
Muscle cramping, shortness of breath secondary to bronchospasm, numbness, and tingling sensations. ↓ Seizures Tetany Refractory hypotension Arrhythmias	No life threatening symptoms (5.0-8.5 mg/dl) ↓ Life threatening symptoms (usually < 5.0 mg/dl)	Mostly require supportive treatment. Oral Supplements for outpatient care. E.g calcium citrate, calcium chloride ↓ IV Fluid replacement, oxygen and serial monitoring ± IV calcium supplement e.g 100-300 mg of elemental calcium over 5-10 minutes: raises the ionised calcium level by 0.5-1.0 mmol/L and lasts for 1-2 hours ↓ Calcium infusion drips (0.5-2.0 mg/kg/h)

EDU — COMMUNITY

MEDICAL WARD

Hypercalcemia

ED specific points of interest

- Extremely rare in patients with eating disorders
- Excess calcium in the diet (>2000 mg/day) can cause hypercalcemia.
- Bed bound for a long period can result in high calcium content in the blood.
- Excess intake of thyroid supplement, thiazide diuretics can result in hypercalcemia.
- Serum parathormone, serum parathormone related protein, Vitamin D level and urine calcium may need to be checked to rule out other causes.
- Acute hypercalcemia needs urgent treatment.
- Mild hypercalcemia (< 3mmol/L) is often asymptomatic.

Management of hypercalcemia

Symptoms	Serum Calcium level	Management	
Tiredness, malaise, dehydration. Suspect hypercalcemia ↓	<3.0 mmol/L [<12 mg/dl] ↓	Mostly require supportive treatment. Serial monitoring of serum calcium levels. Maintaining adequate hydration. Rule out other causes. ↓	C O M M U N I
Renal colic, polyuria, haematuria, Hypertension, bone pain, abdominal pain, ↓	3.0—3.5 mmol/L ↓	Rehydrate with at least 4-6 L of 0.9 % Saline on the first day then reduce to 3-4 L /day for as long as necessary ± IV Biphosphonates treatment of choice when the cause is not known. e.g Pamidronate 60—90 mg IV in 0.9% saline or dextrose over 2– 4 hours, adjusting the rate as necessary	M E D I C A L
Severe vomiting, altered consciousness	> 3.5 mmol/L	± Prednisolone 30-60 mg/ day— only effective in some cases ± Calcitonin/ ? oral phosphate	W A R D

Phosphate

Role of Phosphate

- Essential in building and repairing bones and teeth.
- Plays role in nerve functioning and muscle contraction
- Main component of ATP—the energy chemical in the body
- Essential in cell communication and activation of Vitamin B complex
- Vital part of cell membrane

Rich in

Nuts, Meat, Fish, Cheese
Soy products and Whole grains

ED specific points of interest

- Phosphate is the charged particle (ion) that contains the mineral phosphorus.
- Phosphate level is closely linked to serum calcium level. Hence measurement of one is always done with the other.
- Children will require 350-550 mg/day and adult requirement is 800-1000 mg/day.
- Measurement of Phosphate varies with dietary intake. It drops soon after eating and raises after about 1-2 hours of eating.

Normal Range 0.7-1.0 mmol/L [2.1-3.1 mg/dl]

Dr. Murali Sekar, Dr. Krishnakumar Muthu

Hypophosphataemia

ED specific points of interest

- Isolated deficiency is extremely rare.

- Medications like diuretics and bisphonates can cause hypophosphataemia*. Patients taking a large amount of Aluminium Hydroxide (antacid) can also develop phosphate deficiency.
- Mild symptoms include loss of appetite and weakness. Long term deficiency leads to osteoporosis.
- Low serum phosphorus occurs during starvation and malabsorption. The clinical conditions like hyperparathyroidism, Vitamin D deficiency, diabetic ketoacidosis, acute alcoholism, severe burns, nasogastric suction, and respiratory alkalosis can cause hypophosphataemia.
- Intravenous glucose administration causes phosphate level to drop. This is due to release of Insulin which drives phosphate into the cells. Glucose and phosphate are deposited as Glycogen in muscle and liver.
- Low phosphate level increases calcium absorption in the intestine but renal excretion of calcium also increases.
- In chronic hypophosphataemia, muscle weakness is not seen.
- Oral sodium phosphate is used to treat chronic hypophosphataemia.

---

* Medication-induced hypophosphatemia: a review. Liamis, G.; Milionis, H.J.; Elisaf, M. Qjm. 103(7):449-459, July 2010.

Management of Hypophosphataemia

Symptoms	Serum Phosphate level	Treatment
Tiredness, malaise, loss of appetite and weakness ↓	0.7-1.0 mmol/L [2.1 – 3.1 mg/dl] ↓	Monitor serum level closely; During refeeding add 500mg tds increasing upto 1g qds ↓
Worsening weakness, palpitations and dyspnoea (symptoms of Congestive cardiac failure) ↓	0.4-0.7 mmol/L [1.2-2.4 mg/dl] ↓	
Seizures, Rhabdomyolysis, Haemolytic anaemia Wernicke–Korsokoff syndrome	< 0.4 mmol/L [<1.2 mg/dl]	IV Phosphate Maximum dose of 9 mmol [28 mg/dl] every 12 hours (rapid correction leads to hypocalcemia)

EDU — COMMUNITY

MED ICAL WARD

Dr. Murali Sekar, Dr. Krishnakumar Muthu

Hyperphosphataemia

ED specific points of interest

- Very rare but seen in eating disorder patients with renal failure
- Phosphate containing enemas can result in hyperphosphataemia.
- Excess phosphate binds with calcium resulting in the formation of calcium phosphate. This process could result in reduction of free calcium in plasma.
- Usually no treatment needed apart from treating the cause e.g stopping the phosphate containing laxative.
- Chronic condition could be treated with phosphate binders (e.g Sevelamer Polymer, Lanthanum Carbonate) and dialysis.

Zinc

Role of Zinc

- Involved in many metabolic pathways often acting as coenzyme
- Essential in the synthesis of RNA and DNA
- Zinc is an essential component of various enzymes like alcohol dehydrogenase, alkaline phosphatase.

Rich in

Lamb, leafy vegetables, crab and beef are very rich sources. Zinc is also found in whole grain, pork, milk and eggs.

ED specific points of interest

- Daily RNI is 9.5 mg for men and 7 mg for women.
- Acrodermatitis Enteropathica is an inherited disorder leading malabsorption of Zinc. This condition provides a model to understand the role of Zinc.
- Zinc plays a role in preventing infections leading to diarrhoeal disease, respiratory tract infections. It also plays a role in improving growth in children.

Normal Levels: Plasma Zinc is not a good way to measure body Zinc level. Instead, thymulin activity is measured at times.

Zinc Deficiency

ED specific points of interest

- Acrodermatitis enteropathica is a rare hereditary or acquired disorder of zinc deficiency. It is characterized by acral and periorificial dermatitis, alopecia, diarrhea and growth retardation. There is at least one case report* of a patient with anorexia nervosa and acrodermatitis enteropathica.
- Night blindness is a common feature of Vitamin A deficiency and this symptom is also a feature of Zinc deficiency. Care should be taken to rule out both causes.
- Children suffering from anorexia may experience stunted growth if the Zinc deficiency is not identified and corrected during the period of growth.
- Care needs to be taken when interpreting Zinc level when Serum protein (albumin and globulin) levels are low.
- Impaired wound healing could be a manifestation of Zinc deficiency.
- Prophylactic treatment with Zinc Gluconate 100-200 mg for two months may be recommended.
- Serum Zinc level may not be abnormal until the level is extremely low.
- Symptom may need to be given priority over the serum level and treatment is warranted even if the level is normal.

---

* Acrodermatitis enteropathica with anorexia nervosa. KIM, Sang Tae; KANG, Jin Seuk; BAEK, Jae Woo; KIM, Tae Kwon; LEE, Jin Woo; JEON, Young Seung; SUH, Kee Suck. Journal of Dermatology. 37(8):726-729, August 2010.

Management of Zinc Deficiency

Weight Loss, Impaired or abnormal taste and/ or smell, dry peeling skin, hair loss, diarrhoea and night blindness	Monitor serum level closely. If symptoms persist, Zinc supplementation is justified. (Zinc Gluconate 100-200 mg/day until the level is normal ± 2– 4 months)	EDU / COMMUNITY
↓	↓	
Evidence for Acanthosis (severe peeling of skin, acute respiratory tract infection, severe diarrhoea, opportunistic fungal infections	Level will be very low and the condition is potentially lethal. Rule out other causes. Treat the complications eg antibiotics, antifungals etc. High dose (Zinc Gluconate upto 400 mg/day) may be required. Once the serum level and symptoms improve dose must be reduced to avoid toxicity.	MEDICAL WARD

Zinc Toxicity

ED specific points of interest

- Very rarely seen in ED patients.
- Zinc toxicity can occur with overzealous treatment with Zinc supplement. i.e >400 mg of Zinc for more than four months.
- Excess Zinc impacts the copper and iron metabolism (ingestion of more than 50 mg / day).
- Intake of 75-300 mg/day causes deficiency of copper.
- In medical practice, this condition is usually seen in a patient who has drunk water stored in a galvanised tank.
- Ingestion of 2 grams or more in a short span of time results in symptoms like nausea, fever and vomiting.

Copper

Role of Copper

- Essential for growth
- Essential part of metallo-enzymes like cytochrome C, nitrous oxide reductase.
- Essential in the synthesis of clotting factor V, VIII.
- Plays a central role in maintaining structural integrity of the nuclear membrane.

Rich in

Shellfish, legumes, cereals and nuts.
Cow's milk is a poor source for copper

ED specific points of interest

- An adult has 80 mg of copper in their body of which 40% found in muscle.
- 90% of copper found in the plasma is bound to caeruloplasmin.
- Recommended daily intake is 1.2 mg.
- Higher levels are found in pregnancy.
- Menke's kinky hair syndrome is a sex linked recessive abnormality is a model for studying copper deficiency.

Normal Range Serum Copper: 11-24.4 µmol/L [70-155 µg/dL]

Copper Deficiency

ED specific points of interest

- In ED patients, this condition is more likely to be seen when patient has total parenteral nutrition.
- Copper deficiency in children with anorexia nervosa can result in growth retardation.
- Blood picture of RBC often shows megaloblastic cells. Care needs to be taken to differentiate Copper deficiency with Vitamin B12 & Folate deficiency.
- Symptoms: Oedema when associated with low albumin level, hair and skin hypopigmentation. Neurological abnormalities can also be a presenting feature.
- Association with high cholesterol, heart disease is still under debate with studies claiming direct involvement and a mere coincidence.

Symptoms, Investigations and Treatment in Eating Disorders

Management of Copper Deficiency

Anaemia, infection, impaired healing of wound and bone lesion ↓ Bleeding diathesis	Check serum level Especially when symptoms of RBC, WBC and platelet deficiency occur together. Treat with Copper Sulphate 1-2 mg/day for three months ↓ Treating complications like severe anaemia, heart failure, acute infections, intracranial bleeding may be necessary. Admission to medical ward; isolation to prevent infection.	 

Copper Toxicity

ED specific points of interest

- Wilson's disease is the medical model to understand copper toxicity.
- Very rare in eating disorder patients.
- Symptoms in acute cases are vomiting, diarrhoea and nausea. Can be fatal.
- Chronic copper toxicity can result in liver cirrhosis.
- Built in safety mechanism removes excess copper by excreting it in the bile and faeces.
- Chelation with Penicillamine (1-1.5 g/day) is necessary to remove the excess copper.

Iron

Role of Iron

- Iron is essential for transport of oxygen and cell respiration.
- Iron contained in the myoglobin helps to store oxygen in the muscle cells.
- It is a key component in the enzymes involved in immunity and energy production.

Rich in

Non-heam iron : cereals fortified with iron; Haem iron: haemoglobin and myoglobin in red meat. Haem iron is better absorbed than non-haem iron.

ED specific points of interest

- Normal daily intake is 15-20 mg of which 10% is absorbed.
- Absorptive capacity is increased in hypoxia and iron deficiency. This mechanism fails if for an example blood loss is more than 100 ml per day.
- There is no physiological mechanism in the body to eliminate excess iron. Iron stored as ferritin in the mucosal cells of the gut could be shed, but this process is not under any physiological control.
- There is a diurnal rhythm in serum iron level; higher levels observed in the morning.
- In an average male, about 20 mg of Iron is incorporated into haemoglobin every day.

- Each day 0.5-1.0 mg is lost in faeces, urine and sweat. During menstruation, 0.5-0.7 mg of iron is lost through menstrual blood.

Normal Range Serum Iron 13-32 Micromol/L [72-138 mg/dl]

Three stages of Iron deficiency

Stage of Iron depletion: Serum ferritin level falls to < 12 microgram/L.

Iron stores are depleted.

Stage of ↓erythropoiesis: reduction in the haemoglobin production.

actual Hb level is normal

↓ ferritin level, serum Fe

Transferrin saturation < 19 %

Stage of anaemia: Haemoglobin levels are in the anaemia range

Picture of microcytosis and hypochromia.

Normal levels: Serum Ferritin 70-700 pmol/L [30-300 ng/L] in males

45-450pmol/L [20-200 ng/L] in females
RBC—Microcytosis (MCV <80 fL);
Hypochromia (<27 pg);
Total Iron binding capacity (TIBC): 45-66 µmol/L [250-370 µg/dL];
Transferrin Saturation: Male 20-50%; Female 15-50% <19% in Iron deficiency anaemia.
Serum Iron: Male (11.6-31.7 µmol/L [65-177 µg/dL]; Female 9.0-30.4 µmol/L [50-170 µg/dL].

Dr. Murali Sekar, Dr. Krishnakumar Muthu

Iron Deficiency

ED specific points of interest

- Iron deficiency is more likely in ED patients who are vegetarians.
- Non-haem iron is converted from ferric to ferrous form by the stomach acid. Repeated vomiting soon after dietary intake may impair the absorption of non-haem iron more seriously than haem iron.
- Iron deficiency leads to inadequate haemoglobin synthesis which in turn leads to reduction in oxygen carrying capacity in red blood cells. This defect manifests as iron deficiency anaemia.
- Poor dietary intake over months to years lead to depletion of iron store and anaemia.
- Iron store will be the first to get low followed by lowering of haemoglobin. Hence haemoglobin level may be normal until the iron level drops to very low. Serum Ferritin should be checked along with serum iron and iron binding capacity.
- Effective treatment yields a Haemoglobin increase of 1g/dL per week.
- Intramuscular injection in a patient with very poor muscle mass warrants caution. Iron stores are replaced faster with non oral preparation.

Symptoms, Investigations and Treatment in Eating Disorders

Management of Iron Deficiency Anaemia

| Symptoms of anaemia e.g fatigue, headache, breathlessness. Specific signs like nail changes, angular stomatitis, smooth glossy tongue due to loss of papillae

↓

Angina, intermittent claudications,

Palpitations

↓

Cardiac failure | Hb Level

80 – 120 g/L

[8.0-12.0 g/dL]

↓

50-80 g/L

[5.0– 8.0 g/dL]

↓

<50 g/L

[< 5.0 g/dL] | Oral Iron. Best preparation is ferrous sulphate 200 mg three times daily (beware of side effects)

↓

? Parenteral Iron especially if there is side effects or poor compliance. Intramuscular Iron Sorbitol 1.5 mg of iron /Kg body weight (very painful injection or Intravenous Iron Dextran or Iron Sucrose

↓

?Blood Transfusion, Management of cardiac failure and other complications |

COMMUNITY

MEDICAL WARD

EDU

Iron Toxicity

ED specific points of interest

- Extremely rare in Eating disorder patients.
- Iron toxicity due to excessive dietary intake will not happen due to built-in mechanism in the absorptive capacity in the gut.
- Iron toxicity happens when a patient over doses with more than one ferrous sulphate tablet (60mg) per their body weight in Kg.
- High doses of Fe supplementation reduces absorption of Zinc and other micronutrients.
- Mild symptoms like nausea, vomiting, abdominal pain and diarrhoea can occur. In severe cases shock, metabolic acidosis, hypotension and coma occurs. Death has been reported.
- Majority of patients will not require specific treatment. If there is shock and coma, treatment with desferrioxamine (15 mg/Kg/h i.v) is necessary.

Selenium

Role of Selenium

Selenium is integral to 30 selenoproteins.
They play roles in

- protecting cells against oxidative damage
- the production of triiodothyronine from thyroxine
- antioxidation and transportation in the cells.

Rich in

Fish, eggs, meat and meat products
Vegetarians and vegans are at particular risk.

ED specific points of interest

- Some parts of the world e.g New Zealand has low selenium level in the soil
- Deficiency is extremely rare; Recommended daily intake : 0.04-0.08 mg
- Symptoms are usually related to muscles e.g weakness, pain
- Low serum level is not diagnostic of selenium deficiency. Hence recognising and treating based on symptoms is very essential.
- Treatment is with oral vitamin tablets containing selenium. Intravenous selenium is rarely used.
- Keshan's disease is a selenium responsive cardiomyopathy seen in China.

Normal Range Whole blood 0.091-0.12 μg /ml

Other Trace Elements

Recommended daily intake

Element	Notes	Recommended daily intake
Cadmium	No documented symptoms	??
		25—30 microgram/day in adults
Chromium	linked to glucose intolerance	??
Cobalt	Essential for Vitamin B12 synthesis	1.5—7.0 mg
Manganese	can lead to osteoporosis. In treatment resistant osteoporosis in patients with anorexia nervosa, consider adding Manganese in the treatment.	50—400 microgram/day for adults
Molybdenum	prolonged TPN may lead to deficiency.	0.5—1.5 microgram/day for children

Fluoride
- ☐ Recommended daily intake is 2-3 mg
- ☐ Rich in sea fish and tea.
- ☐ Drinking water is a rich source of Fluoride. If level is <0.7 mg/L, dental caries results.
- ☐ Increased level of fluoride results in pitting of teeth and discoloration.

Iodine
- ☐ Iodine is a component of T3 and T4.and is required for growth.
- ☐ Recommended daily intake is 140 microgram.
- ☐ Rich in milk, meat and sea foods.
- ☐ Iodine deficiency is a very common mineral deficiency.
- ☐ Iodine deficiency leads to symptoms of hypothyroidism.

Vitamins

Introduction

It was suggested that, when plants and animals began to move from the sea to rivers and land about 500 million years ago, environmental deficiency of antioxidants was a challenge to the existence of terrestrial life. Terrestrial plants slowly optimized the production of "new" endogenous antioxidants such as ascorbic acid (Vitamin C), polyphenols, flavonoids, tocopherols, etc. The terrestrial animals have not mastered this art of production of antioxidants. Hence they require the vitamins to be supplied from outside through diet.

The name 'vitamin' stands for 'vital amines'. As it denotes, vitamins were once thought to be amines. Now we know that vitamins are a group of compounds with a wide range of chemical base. The current understanding is that a "vitamin" refers to a number of 'vitamer' compounds. These compounds show the biological activity associated with a particular vitamin e.g Vitamin A has retinol, retinal and four carotenoids as its 'vitamers'.

There are two principal groups:

Fat soluble: Vitamin A, E, D and K which can be stored in the body

Water soluble: Rest of the vitamins are as water soluble (e.g) Vitamin B compound, Vitamin C.

Vitamins have diverse biochemical roles. It include hormone like functions in regulating mineral metabolism (Vitamin D), regulator of cell growth and differentiation (Vitamin A), antioxidant (Vitamin E and C), enzyme co-factors (Vitamin B compounds) and co-enzymes (Folic acid).

Humans must consume vitamins periodically to avoid deficiency. Human bodily stores for different vitamins vary widely; vitamins A, D, and B12 are in significant amounts in the human body, mainly in the liver, and an adult human's diet may be deficient in vitamins A and D for several months and B12 in some cases for years, before developing a deficiency condition. However, niacin and niacinamide are not found in human body in significant amounts so stores may only last a couple of weeks. For Vitamin C, the symptoms of deficiency can start between one to six months.

Deficiencies of vitamins are as either primary or secondary. A primary deficiency occurs when an organism does not get enough of the vitamin in its food. A secondary deficiency may be due to an underlying disorder that prevents or limits the absorption or use of the vitamin, due to a "lifestyle factor", such as smoking, excessive alcohol consumption, or the use of medications that interfere with the absorption or use of vitamin. Hence the vitamin deficiency observed in sufferers of eating disorders will fall under the 'primary deficiency' category.

In the nomenclature of vitamins, we can observe that the set of vitamins skips directly from E to K. It is because the vitamins corresponding to letters F-J wereeither over time (Vitamin F as essential fatty acids), discarded as false leads (Vitamin O as this can be synthesised in the body), or renamed because of their relationship to vitamin B, which became a complex of vitamins (Vitamin G as Riboflavin and Vitamin H as Biotin).

Water Soluble Vitamins

Vitamin B12

Role of Vitamin B12

- Main function is methylation of homocysteine to methionine
- Myelination of nerve fibres
- Metabolism of folate coenzymes

Rich in

Meat, fish, eggs and milk. Not rich in plant foods.

ED specific points of interest

- Average daily intake is 5-30 micrograms of which 2-3 microgram is absorbed
- Average adult body has a reserve of 2-3 mg. Poor dietary intake can lead to exhaustion of reserve over two years.
- Deficiency more commonly seen in vegans
- 1-2 mg of oral Vitamin B12 is absorbed by simple diffusion. Therefore, intrinsic factor is not always necessary.
- Hypokalemia is a likely complication especially if the patient is vomiting or using laxatives. Iron deficiency and hyperuricemia are also likely to occur in the early stage of treatment.
- Blood picture improves within 48 hours. Improvement in neurological symptom may occur over 6-12 months.

- Treatment with just folic acid will provide a haematological response, but neurological condition will worsen.

Normal Range: 150-590 pmol/L [200-800 pg/mL]

Vitamin B12

ED specific points of interest (contd.,)

- Cyanocobalamin is a commercially available from which is converted to natural form of cobalamin.
- Salivary enzyme haptocorrin and intrinsic factor secreted in the stomach are essential for absorption. Some patients who regularly chew and spit food may be more prone to develop this deficiency due to reduced availability of haptocorrin
- Few patients exhibit symptom as cobalamin is manufactured in intestine. Regular use of laxative could affect this function hence close monitoring may be needed in this group of patients.
- No human toxicity is reported.

Management of Vitamin B12 Deficiency

Symptoms	Level	Treatment	
Symptoms of anaemia e.g fatigue, headache, breathlessness. Specific signs like angular stomatitis, red sore tongue	150- 60 pmol/L [80 -200 pg/mL]	Oral Vitamin B12 tablets 2 mg / day	EDU COMMUNITY
↓	↓	↓	
Neurological changes (if left untreated can lead to irreversible changes) Polyneuropathy, posterior then lateral columns of spinal cord (Subacute combined degeneration). Symptoms are symmetrical paraesthesiae in fingers and toes, loss of vibration sense and proprioception, ataxia, dementia and psychiatric symptoms.	<60 pmol/L [< 80 pg/dL]	Hydroxocobalamin 1 mg Intrmuscular injection; 5-6 injections over three weeks. Repeat Hydroxocobalamin 1 mg i.m every three months. Treatment with both Vitamin B12 and Folic Acid may be necessary in severe deficiency even if the folic acid level is normal.	MEDICAL WARD

Folic Acid

Role of Folic Acid

- Active metabolites of folate (folate polyglutamates) play an essential role in amino acid metabolism and DNA synthesis through its participation in number of single carbon transfer reactions.
- Essential in the formation of methionine along with Vitamin B12.

Rich in

Green vegetables e.g spinach, broccoli, Brussels sprout, offal such as liver and kidney, most fruits, potatoes, milk, eggs, oats, brown rice contain a moderate amount of folate.

ED specific points of interest

- Cooking results in loss of 60-90 % of folate
- Red blood cell folate is a better measure than serum folate level.
- Daily requirement is about 100 microgram.
- Normally human body has a store of 10 mg. If nutritional intake stops, deficiency can occur within four months.
- Number of drugs e.g anticonvulsants, antibacterial agents affect folate metabolism resulting in deficiency
- Excess folate in the body is extremely rare

Normal Range: Serum Folate 12.2-90 nmol/L [5.4-40.0 ng/mL] Red cell Folate 340-1020 nmol/L/cells [150-450 ng/mL/cells]

Patients may be asymptomatic. Symptoms of anaemia e.g fatigue, headache, breathlessness can occur. No neurological symptom is associated with folic acid deficiency.	Folic Acid 5 mg/ day for four months.

Role of Thiamine

- Thiamine is needed for the metabolism of fat, carbohydrate and alcohol.
- A coenzyme thiamine pyrophosphates an essential cofactor in citric acid cycle, catabolism of branch chain amino acids.
- Also plays a role in hexose monophosphate shunt.

Rich in

Cereal products, pulses, nuts, milk and some vegetables.
In UK bread flour is with thiamine, and this is a requirement by law.

ED specific points of interest

- Daily requirement depends on calorie intake; high carbohydrate intake requires a high amount of thiamine.
- Red cell transketalose is used to measure thiamine status in the body. Deficiency is by an increase in the enzyme's activity. A typical increase of 70-100% is seen in Wernicke's encephalopathy.
- Recommended average daily intake is between 1.5-2.0 mg.
- Chronic intake of >3 mg/day can lead to toxicity. The symptoms are headache, irritability, weakness, tachycardia, ataxia and pruritus.

Thiamine Deficiency

- Body reserves are remarkably small hence signs of deficiency occurs rapidly with inadequate dietary intake. Complete deprivation of thiamine in the diet can lead to deficiency state in 18 days.
- Low magnesium and phosphate commonly observed in thiamine deficiency.
- One of the common vitamin deficiency observed in patients with anorexia nervosa. Care to be taken to avoid this during refeeding.
- Dry and wet beriberi rarely occur together.
- Wernicke-Korsakoff syndrome is due to ischemic damage to the brainstem and its connections. Symptoms and signs are:
- Eye signs: nystagmus, bilateral lateral recti palsy, conjugate eye palsy, fixed pupils and papilloedema.
- Brain stem signs: Ataxia, cerebellar signs and vestibular paralysis.
- Cognitive deficits: loss of memory, confabulation, restlessness, stupor and coma.
- Untreated Wernicke-Korsakoff syndrome leads to irreversible brain damage.
- Transketalose test is not widely available; therefore, we need to rely on symptoms and signs. In doubt start the treatment. Anaphylaxis has been reported.

Management of dry Beri-Beri

Heaviness and stiffness of legs	Thiamin 50 mg I.V or I.M for three days followed by 100 mg orally everyday for three months.	COMMUNITY
↓	↓	
Weakness, numbness, loss of ankle jerk followed by involvement of arms and trunk.	May need admission. Prevention of fall and resultant injury.	EDU
↓	↓	
Cerebral involvement producing Wernicke-Korsakoff syndrome (can occur in the acute stages too)	Wernicke– Korsakoff is a medical emergency. Admission to a specialist unit is warranted	MEDICAL WARD

Mangement of wet Beri beri

Oedema in legs with low vitamin level. Check for other causes.	Thiamin 50 mg I.V or I.M for three days followed by 100 mg orally everyday for three months.	EDU
↓	↓	
Oedema extending to whole body, ascites, pleural effusion	May need admission to medical ward. Close monitoring of fluid balance and respiratory failure.	MEDICAL WARD
↓	↓	
Heart failure (high cardiac output state), arrhythmia due to conduction defect.	Admission to a cardiac unit and management of cardiac failure.	

Management of Wernicke-Korsakoff Syndrome

Acute stage is a medical emergency requiring urgent treatment.

Requires thiamine 250 mg i.m or i.v infusion once daily for three days.

Other vitamins are usually added to the regime as oral medication.

Beware of anaphylaxis.

Thiamine 100 mg should be continued for three months, may be longer in some cases.

Riboflavin (Vitamin B2)

Role of Riboflavin

- Promotion of normal cell growth
- Synthesis of steroids, glycogen and red blood cells
- Important for Iron absorption
- Helps in maintaining mucosal surface

Rich in

Eggs, Milk, Liver, Yeast and Fortified cereals.

ED specific points of interest

- Erythrocyte Glutathione Reductase Actvity* is the method of choice of assessing riboflavin status in the body as this reflects the tissue saturation and the long term picture.
- Riboflavin is unstable in the sunlight. For example, 70% of riboflavin content will be lost from milk if it is exposed to sunlight for more than four hours.
- Toxicity is extremely rare as absorption in the G.I tract is limited.

Normal Range: Normal when EGRA* activity coefficient is <1.2 Intermediate 1.2-1.4 and low if EGRA is > 1.4.

Riboflavin Deficiency

- Severe deficiency is very rare in UK but is reported in patients suffering from anorexia nervosa.
- Lesions in the mucosal surface can manifest as angular stomatitis, cheilosis, magenta tongue and lesions on the mucosal surface of genitalia.
- Seborrheic skin lesions and corneal vascularisation are also seen in Riboflavin deficiency.
- Deficiency is often seen with other nutrient deficiencies.

Management of Riboflavin Deficiency

Riboflavin 5 mg usually as Vitamin B complex tablets.

Niacin (Vitamin B3)

Role of Niacin

- Nicotinamide Adenine Dinucleotide (NAD) a metabolite of Niacin is essential for oxidative steps in energy production.
- Nicotinamide adenine dinucleotide phosphate is necessary in fatty acid synthesis.
- Also plays a role in tissue respiration and detoxification.

Rich in

Beef, pork, chicken, eggs, milk, wheat and maize.

ED specific points of interest

- Niacin can be synthesised in humans from tryptophan. Approximately 50% of niacin is synthesised in the human body. 1 mg of Niacin is produced from 60 mg of tryptophan. For this conversion Vitamin B6 and riboflavin are essential. Hence deficiency of these vitamins can also cause pellagra.
- N'—methylnicotinamide is a metabolite of niacin. This assay requires 24 hour urine collection and the test is not widely available.
- Amount of niacin in food is usually measured as niacin equivalent (NE). Recommended intake is 6.6 mg of niacin equivalent (NE)/1000Kcal/day.

- Doses of 1-2 g/day are used in the treatment of hypertriglyceridaemia and hypercholesterolaemia. Larger doses (3-6g/day) cause impairment in liver function and glucose metabolism.

Normal Range: N'-methylnicotinamide:creatinine ratio >1.5 mmol/L

ED specific points of interest

- The classical features are dermatitis, diarrhoea and dementia. But these changes are not always seen. Dermatitis: ulceration, thickening, dryness and pigmentation in the areas of skin exposed to sunlight. Diarrhoea: is the common symptom, but constipation can be a presenting symptom. Other symptoms like glossitis and angular stomatitis frequently occur.
- Dementia: this occurs in chronic condition. Depression, apathy and thought abnormalities occur. Hallucinations and encephalopathy can also occur.
- Niacin deficiency is likely to occur in severely restricting patients who also abuse alcohol.
- Care should be taken to avoid mistaking pellagra-like skin lesions observed rarely in patients with anorexia nervosa*.

---

* Pellagra-like erythema on sun-exposed skin of patients with anorexia nervosa.
Sato, Mami 1; Matsumura, Yumi 1; Kojima, Ayako 1; Nakashima, Chisa 1; Katoh, Mayumi 1; Kore-Eda, Satoshi 1; Miyachi, Yoshiki 1Journal of Dermatology. 38(10):1037-1040, October

Management of Pellagra

Skin changes and diarrhoea	Acute	Loading dose of Nicotinamide 300 mg followed by 50 mg/day results in dramatic improvement. Severe cases need admission.	MEDICAL WARD	COMMUNITY
↓	↓	↓		
Symptoms of depression, mania, hallucinations	Chronic	Admission to a medical/ psychiatric ward may be necessary as is the treatment with psychotropic medication. But the main stay is treatment with Nicotinamide.		

Vitamin B6 (Pyridoxine, Pyridoxal, Pyridoxamide)

Role of Vitamin B6

- Synthesis of non essential amino acids
- Synthesis of biologically active amines e.g adrenaline, noradrenaline, GABA, serotonin and histamine
- Synthesis of porphyrins including haemoglobin
- Metabolism of hormones
- Conversion of glycogen to glucose in muscles and tryptophan to niacin

Rich in

Meat, Cereals, bananas and pulses

ED specific points of interest

- Severe deficiency is extremely rare.
- Clinical signs like inflammation of the tongue, neuropathy could be associated with Vitamin B6 deficiency. These signs occur with deficiency of other vitamins as well hence correction of other vitamins like Thiamine is essential.
- Poor haem synthesis could lead to anaemia.
- Toxicity can occur if daily intake exceed 500 mg/day. The recommended dose is 10 mg/day.

Normal Range Plasma Pyridoxal Phosphate 30-140 nmol/L [5-24 ng/mL]

Management of Vitamin B6 Deficiency

Vitamin B6 deficiency is extremely rare as mentioned before.

Treatment of co occurring vitamin and mineral deficiency is essential.

Multi-Vitamin preparations often contain enough Vitamin B6 hence a special Vitamin B6 preparation is not necessary.

Biotin (Vitamin H)

Role of Biotin

- Synthesis and metabolism of fatty acids
- Metabolism of branched chain amino acids
- Gluconeogenesis.

Rich in

Milk, Liver, Kidney, Eggs and other dairy products.

ED specific points of interest

- Biotin is synthesised by bacteria in the colon. Hence patients who are abusing laxatives might have an increased risk of Biotin deficiency. This is especially the case if the patient restrict his/her dietary intake too.
- Biotin deficiency can be produced when a large amount of raw egg white is consumed. The raw egg white contains avidin which denatures Biotin.
- Daily intake of 10-200 microgram is generally considered adequate.
- Deficiency usually manifests as dermatitis, glossitis, hair loss and hypercholesterolaemia.
- No reports of biotin toxicity.

Normal Range Blood level of Biotin 0.22-0.75 microgram/ml

Management of Biotin Deficiency

- Biotin deficiency can lead to acrodermatitis a skin condition characterised by periorificial or acral dermatitis*. Acrodermatitis is also seen in zinc deficiency, essential fat acid deficiency and a protein energy malnutrition called kwashiorkor.
- Treatment with Biotin 10-200 microgram/day is usually sufficient.
- Multi-Vitamin preparations usually contain enough Biotin. Therefore, a special Biotin preparation is not required.

* Acrodermatitis Gehrig, Kathryn A; Dinulos, James GH. Current Opinion in Pediatrics. 22(1):107-112, February 2010.

Pantothenic Acid

Role of Pantothenic Acid

- It is involved in carbohydrate and lipid metabolism.
- Essential component of Coenzyme A.

Rich in

Meat, eggs, green vegetables, yeast and peanuts.

ED specific points of interest

- Clinical deficiency has been reported with symptoms of fatigue, apathy, numbness, cramps, symptoms of hypoglycaemia, and sleep difficulties.
- Deficiency was produced in an artificial feeding setting, in the laboratory. Symptoms of such state include burning of feet, muscle weakness, vomiting etc.
- 3-7 mg/day is considered adequate.

Normal Range

Blood level >100 microgram/ml; Urine Excretion 1-15 mg/day

Vitamin C (Ascorbic Acid)

Role of Vitamin C

- Synthesis of collagen and bile.
- Synthesis of noradrenaline from dopamine.
- Essential in the metabolism of many drugs and carcinogens in the liver.
- Improves absorption of Iron when consumed in the food together.
- Hormones and releasing factors are activated by Vitamin C

Rich in

Citrus fruits, other fruits like mango, papaya, sweet potato, pepper and broccoli

ED specific points of interest

- A normal exchangeable body pool is about 900 mg (5.1 mmol)
- Humans and few other species in the animal kingdom do not have the capacity to synthesize ascorbic acid from glucose.
- Ascorbic acid is highly unstable. It is leached out of vegetables when placed in water, oxidised during cooking or lost from potatoes during storage.
- High dose of ascorbic acid (1-2 gram/day) is often taken as a remedy for the common cold. Sudden cessation may lead to rebound scurvy and continuation of such a heavy dose over a period may result in diarrhoea and kidney stone formation.

- Daily requirement is 40 mg, but smoking accelerates the metabolism of Vitamin C hence smokers require 80 mg.
- Plasma levels are more practical to measure than the leucocyte ascorbate measurement.

Normal Range Plasma ascorbic acid 17-114 µmol/L [0.3-2.0 mg/dL]; Leucocyte ascorbate 1.1-2.8 pmol/L/10^6 cells

Vitamin C Deficiency (Scurvy)

ED specific points of interest

Symptoms

- Swollen, spongy gums with a tendency to bleed easily with no or minimal injury.
- Common symptoms are: Keratosis of hair with corkscrew appearance
- Spontaneous haemorrhages and bruising\
- Scurvy is often associated with anaemia and failure of wound healing. Anaemia may be of Iron deficiency or folate deficiency in nature due to poor absorption or availability of these nutrients.
- Symptoms are mainly due to abnormalities of connective tissue i.e failure of synthesis to collagen.

Management of Scurvy

Symptoms	Level	Management
No symptoms	>17 µmol/L [>0.3 mg/dL]	If symptomatic, rule out other causes
↓		↓
Non-specific weakness, muscle pain	11-17 µmol/L [0.2-0.3 mg/dL]	Loading dose of 250 mg of ascorbic acid daily until level comes to normal range followed by 40 mg daily until nutrition improves
↓	↓	↓
Anaemia, spontaneous bruising, failure of wound healing. Symptoms like spontaneous haemorrhage necessitates admission.	<11 µmol/L [<0.2 mg/dL]	As above and admission to relevant medical wards if haemorrhage and cardiac compensation are present

COMMUNITY — EDU — MEDICAL WARD

Fat Soluble Vitamins

Vitamin A

Role of Vitamin A

- Retinaldehyde is found in the rods (rhodopsin) and cones (iodopsin) of the retina.
- Retinal and retinoic acid are involved in the control of cell proliferation and cell differentiation
- Retinyl phosphate is a cofactor in the synthesis of glycoproteins
- Also plays a role in growth, embryogenesis and immune response

Rich in

Vitamin A (Retinol): Liver, milk, butter, cheese, egg yolk and fish oils.

Beta-carotene: Green vegetables, carrots and other yellow and red fruits.

ED specific points of interest

- Vitamin A activity is usually measured as retinol equivalent.
- Fat is necessary for absorption of Vitamin A. As retinol is found in animal products the fat present in the food should help the absorption. However in some of the patients who significantly restrict fat intake absorption of Vitamin A would be reduced.
- Nutritionally 6 micrograms of beta carotene is equivalent to 1 microgram of preformed retinol.

- Daily requirement is 600-700 microgram of Retinol Equivalent per day.

Normal Range Plasma level of Vitamin A 1.05-3.32 µmol/L [30-95 µg/dL]

Vitamin A Deficiency

ED specific points of interest

- Protein deficiency (retinol binding protein—a globulin) can cause vitamin A deficiency even though the body stores of vitamin A is normal.
- Eye signs: Night blindness, conjunctival xerosis, bitot's spots, corneal xerosis, and corneal softening (Keratomalacia)
- Skin keratinisation and hyperkeratosis of sebaceous skin glands can occur.
- Vitamin A deficiency can cause nutritional deficiency and low immunity. This will make the individual more susceptible to respiratory and gastrointestinal infections.
- Vomiting and laxative abuse may inhibit absorption of oral vitamin A.

Management of Vitamin A deficiency

No symptoms	> 1.05 µmol/L [>30 µg/dL]	If symptomatic (for example Bitot's spots can occur without Vitamin A deficiency) rule out other causes
↓	↓	↓
Marginal ± Symptoms	0.3-1.0 µmol/L [10-30 µg/dL]	Dietary intake should be increased
↓	↓	↓
Eye changes, skin changes, Lowered immune response and nutritional anaemia	< 0.3 µmol/L [<10 µg/dL]	Retinol Palmitate 30 mg orally for 2 days. Vitamin A 30 mg i.m if diarrhoea and vomiting is present. Treatment of superadded infection and referral to Ophthalmologist

COMMUNITY

MEDICAL WARD

Vitamin A Toxicity

ED specific points of interest

- Even single doses of vitamin A of more than 300 mg in adults and 100 mg in children can cause toxicity.
- Acute excess vitamin A causes bone damage, hair loss, double vision, vomiting and headache. Chronic vitamin A excess (>10mg of total vitamin A in the body) causes headache, muscle and bone pain. This may also cause ataxia, skin disorders, alopecia, skin disorders and liver toxicity.
- Vitamin A is associated with birth defects. Pregnant women are advised to take <3 mg /day.
- No symptom is associated with excess carotenoids in food apart from yellowish discoloration of the skin.

Vitamin D

Role of Vitamin D

- Maintains plasma calcium by regulating calcium absorption and excretion
- Bone mineralisation
- Helps immune responses
- May inhibit cell proliferation in cancer cells.

Rich in

Cod liver oil, oily fish, milk, cereals, eggs and liver

ED specific points of interest

- Only vitamin to be affected by a season of the year i.e higher levels in summer and lower during winter months.
- Average daily production is between 3-4 microgram/day.
- Usually dietary intake is not necessary since the production happens during exposure to sunlight.
- Toxicity is not seen as excessive Vitamin D production does not usually occur, and excess supplemental intake are rapidly metabolised and eliminated. In theory, hypervitaminosis D can result in hypercalcemia.

Normal Range: Serum 25-hydroxyvitamin D > 25-137 nmol/L [10-55 ng/mL]

Formation of Vitamin D

7-dehydrocholesterol

(skin)

↓ Sunlight

Cholecalciferol

↓ in Liver

25-hydroxyvitamin D

↓ in Kidney

(regulated by Parathormone, Phosphate)

↓

1, 25—hydroxyvitamin D

Vitamin D Deficiency

ED specific points of interest

- Deficiency is common in patients who stay indoors and are fully covered.
- Malabsorption of ergocalciferol can lead to Vitamin D deficiency.
- Plasma calcium and phosphate level decline in severe deficiency state.
- Alkaline Phosphatase level is elevated in mild and severe cases.
- Symptoms are usually due to osteomalacia (in adults) characterised by bone pain, muscle weakness and rickets (in children) characterised by bone deformity, bone pain and muscle pain.

Management of Vitamin D deficiency

To prevent osteoporosis: Calcium 700-1000mg/day + Vitamin D 400-800 IU.

To treat Rickets: Vitamin D 400-800 IU

Rarely IV preparation may be necessary.

Vitamin E

Role of Vitamin E

- Antioxidant
- Contributes to biological membrane stability
- Protects cell structures
- Affects cell proliferation and growth.
- Regulates prostaglandin synthesis.

Rich in

Vegetable and seed oils, cereals and nuts. Animal food is a poor source.

ED specific points of interest

- Vitamin E has eight naturally occurring forms of which alpha tocopherols are the most active forms.
- Recommended daily intake 7-10 mg/day.
- Plasma or serum level should be corrected for plasma lipid level i.e so many milligrams of Vitamin E per milligram of plasma lipid.
- Vitamin E is linked to prevention of ischaemic heart disease, peripheral vascular disease, occurrence of cancer and Alzheimer's disease. The evidence is not clear.

Normal Range: Plasma levels between 12-46 µmol/L [5-20 µg/mL]

Vitamin E and Polyunsaturated Fatty Acid (PUFA)

Vitamin E intake, requirement and absorption are closely associated with lipids (PUFA) in the diet. Requirement for Vitamin E is determined by PUFA content in the diet. High PUFA Intake requires high Vitamin E. Estimated value is that 0.4 mg of alpha tocopherol is required per gram intake of PUFA.

Vitamin E Deficiency

ED specific points of interest

- Since the absorption and requirement are determined by fat in the diet, eating disorder patients who restrict fat intake are more prone to develop a deficiency.
- Symptoms of deficiency are mainly neurological. Reduced tendon reflexes, loss of touch and pain sensations, unsteady gait, impaired eye movement and incoordination are the main symptoms.

Management of Vitamin E deficiency

Mild symptoms are usually managed by improving the dietary intake of PUFA and Vitamin E rich foods.

Moderate to severe deficiency may need Vitamin E supplementation (5-20 IU/day) or Vitamin E injections (usually reserved for severe ataxia)

Vitamin E Toxicity

It is extremely rare but happens when >900 mg/Kg of the diet can produce symptoms like headache, nausea, double vision and GI disturbances. May also affect creatinine metabolism. At very high level, Vitamin E can antagonise Vitamins A, D and K leading to their deficiency

Vitamin K

Role of Vitamin K

- Essential for synthesis of clotting factors II, VII, IX and X.
- Osteocalcin a bone protein requires Vitamin K for its synthesis.

Rich in

Green leafy vegetables, milk, egg and vegetable oils.

ED specific points of interest

- Two naturally occurring forms. Vitamin K1 is found in plants. Vitamin K2 is synthesised by intestinal flora.
- Vitamin K level is measured by coagulation factors that are dependent on Vitamin K for their production.
- Approximate daily requirement is between 0.5-1.0 microgram/ Kg/day.
- Vitamin K requires bile salts. Any obstruction e.g Cholestatic Jaundice can lead to reduction in the absorption of Vitamin K.

Normal Range

Clotting factor levels are usually used as substitute measure.

Bleeding time	3-9.5 Min
Prothrombin time	10-13 sec
Partial thromboplastin time (activated)	22-37 sec

Vitamin K Deficiency

ED specific points of interest

- Chronic laxative abuse, use of antibiotics for prolonged periods can lead to deficiency in eating disorder patients.

Management of Vitamin K deficiency

Acute presentation i.e haemorrhage should be treated with Phytomenadione injection (10mg). Follow up with oral therapy (Menadiol sodium phosphate 10 mg/day) will be usually advisable.

Renal Function Tests

Introduction

The kidneys are paired organs, 11-14 cms in length, 5-6 cm in width and 3-4 cm in thickness. It has two parts namely cortex (outer part) and medulla (inner part). The functional unit of the kidney is the 'nephrons' and each nephron consists of glomerulus, proximal convoluted tubule, loop of Henle, distal convoluted tubule and collecting duct. The filtrate from the glomerulus passes through the abovementioned parts of the nephron in that order. The kidney is supplied by renal artery, which subdivides to a number minute arterioles. There are afferent arterioles (which takes the unfiltered blood to the capillary bed of the glomerulus) and efferent arterioles (which collects the blood after its contents were filtered by the glomerulus).

Glomerulus is a basket like structure which represents the head of the nephron. The blood from the afferent arteriole passes through the sieve or filter formed by endothelial cells, basement membrane and epithelial cells.

Functions of the Kidneys:

Every minute about 25% of blood pumped out by heart goes through the two million glomeruli in the kidneys. A pressure difference of approximately 10 mm Hg drives the fluid in the blood within the arterioles into the glomerulus and proximal convoluted tubule. Every day roughly about 170-180 L of this filtrate is formed. Most of this will reabsorbed (60-80 % of filtered water and sodium will be reabsorbed in proximal convoluted tubule along with almost all potassium, bicarbonate, glucose and amino acids. The rest of the water is reabsorbed in distal convoluted tubule and collecting ducts by the influence of aldosterone and antidiuretic hormone. Hence the final urine output is normally between 1-2 L / day.

Various condition can alter the amount of water excreted, retention of minerals, proteins etc. For example, potassium is freely filtered at the glomerulus, almost completely reabsorbed in the proximal convoluted tubule but secreted in the distal tubule and collecting ducts in varying degrees. More detail could be found in the acid-base balance section.

Glomerular Filtration Rate

- Glomerulus are the basket shaped part of the nephron which is the functional cell of Kidneys.
- Kidneys main function of eliminating toxic wastes is started here in the glomerulus with those chemicals being filtered off the blood stream.
- Rate at which these chemicals are filtered and excreted is very essential for the prevention of accumulation of these chemicals in the body.
- This is measured by the Glomerular Filtration Rate (GFR). In clinical practice, this is done by measuring the creatinine clearance (and NOT Creatinine Excretion) from the body.

The reason being

- Creatinine is one of the waste material produced in the body which the kidneys need to excrete.
- Production of Creatinine is constant and not influenced by diet but internal factors like age, sex and muscle mass. Hence measurement of Creatinine is more accurate measure of GFR than other waste materials e.g Urea.
- Creatinine excretion also depends on another important component of the renal function—Tubular secretion, but this is relatively small. Hence measurement of the amount of

Creatinine excreted will give good enough account of Glomerular filtration rate in normal or near normal renal function.

Serum Urea and Creatinine levels do not raise above the normal level until the GFR capacity of the kidneys fall below 50-60%. Hence measurement of Urea and Creatinine without measuring the correspondent GFR can lead to false assumption of healthy status whilst the Kidneys filtration capacity is compromised. Moreover early identification of reduction in GFR could inform early interventions.

Measurement of GFR

- From the measurement of Creatinine clearance (Creatinine clearance = U x V/P U is concentration of creatinine in urine collected over 24 hours; V is the rate of urine flow in mL/min; P = plasma concentration of creatinine).
- Other techniques to measure GFR:

    - Injection of compounds e.g 125 I iothalamate and serial measurement of their plasma concentration. The GFR may be calculated from the exponential fall in the blood level.
    - Cystatin C is an enzyme produced in the body. It is filtered by glomerulus and not secreted in the tubules. Hence measuring the urinary concentration of this enzyme is a good way to measure GFR. This test is highly sensitive so even mild renal impairment could be identified (when the creatinine is still normal). This test is not widely available.
    - Calculated GFR—Using special formulae:

Cockroft—Gault equation

Modification of Diet in renal disease (MDRD) equation—This method takes into account various factors like age, sex, creatinine and ethnicity.

- A study* reported that Cockroft—Gault equation is not particularly sensitive in patients with extremes of weight whereas MDRD may underestimate the renal function in normal-high GFR. Mayo clinic quadratic (MAYO) seems to be the method of choice.

* Estimation of Renal Function in Patients with Eating Disorders. Fabbian, Fabio MD 1, *; Pala, Marco MD 1; Scanelli, Giovanni MD 2; Manzato, Emilia MD 2; Longhini, Carlo MD 1; Portaluppi, Francesco MD 1, 3 International Journal Of Eating Disorders. 44(3):233-237, April 2011.

Creatinine

- Creatinine is synthesised from Creatine (phosphorylcreatine)
- Creatine is synthesised in Liver, whereas creatinine is formed in skeletal muscle.
- Creatine is not usually found in urine of men but present in children, women during and after pregnancy and occasionally in non-pregnant women.
- Any excess muscle breakdown including vigorous exercise can lead to creatinuria. Starvation alone can lead to excess creatinine in urine.

Normal plasma creatinine Level: 53-133 µmol/l [0.6-1.5 mg/dL]

Synthesis of Creatinine

Protein in the diet

↓

Broken down to

Methionine, glycine and arginine   Intestine

(amino acids)

↓

Creatine   Liver

↓

Phosphorylcreatine   Muscle

↓

Creatinine   Kidney

Urea

- Urea is formed from toxic ammonia ions of amino acids.

- Normal Urea level (Serum blood urea nitrogen BUN) is 2.9-8.9 mmol/L [7-30 mg/dL]

- Urea is formed in Liver and excreted in urine.

- In severe liver disease urea level falls (Blood urea nitrogen [BUN]) but ammonia level in blood raises leading to toxic effects.

Synthesis of Urea

Protein in food

↓

Amino acids   Intestine

↓

NH4 ions   Liver

↓

Citrulline   Mitochondria
(Liver)

↓

Urea + Arginine

↓

Urea excreted in Urine

Acute Renal Failure

- Depression of glomerular filtration rate is the main cause leading to renal failure.
- Altered regulation of salt and water balance could be a result as well as cause for renal failure in eating disorder patients.
- Low volume of blood and low blood pressure lead to reduction in perfusion to kidneys. This will lead to prerenal uraemia.
- Low water intake, sudden cessation of water loading can upset the intrinsic autoregulation mechanism in the kidneys leading to acute renal failure.
- Usually present as reduced urine output. Serum Creatinine and urine output are used to classify the condition as mild, moderate and severe.
- Three types of renal failures namely acute, acute on chronic and chronic renal failures have different causes, treatment options and prognosis.
- Vigorous exercise—myoglobinuria but not creatinuria leading to ARF
- Beware of the combination of reduced urine output with raised serum creatinine

Management of Acute Renal Failure

Serum Creatinine 1.5 times the normal	Urine output <0.5 ml/Kg/Hour for 6 hours		M E D I C A L
Serum Creatinine 2 times the normal	Urine output <0.5 ml/Kg/Hour for 12 hours	Acute Renal failure is a medical emergency. Any evidence for increase in the serum creatinine with reduced urine output should raise suspicion of renal failure.	
Sserum creatinine 3 times normal or Serum creatinine > 350 micromol/L with an acute rise of creatinine of atleast 40 micromol/L	Urine output <0.3 ml/Kg/Hour for 24 hours	In hypovolaemia and hypotension associated with eating disorders, treatment should be concentrated on the fluid replacement.	W A R D

Chronic Renal Failure

- Eating disorders as such will not lead to Chronic renal failure.
- Common causes for chronic renal failure are congenital conditions like polycystic kidney disease, glomerular disease, vascular disease e.g reno-vascular disease, urinary tract obstruction due to various causes.
- In a patient already suffers from any of the above are the high risk patients to develop acute on chronic renal failure if they indulge fluid restriction etc.
- Renal osteodystrophy could worsen or initiate osteoporotic changes in AN patients.
- Chronic kidney disease is associated with the loss of cyclical changes in sex hormones leading to oligomenorrhoea or amenorrhoea. Hence patients can lose their menstruation despite being in a health weight range.
- Hyperkalemia is seen either as ? Intrinsic to CRF or secondary to use of potassium sparing diuretic.
- Chronic potassium depletion affects the proximal convoluted tubule and interstitium resulting in reduction of glomerular filtration. This could lead to renal failure*. Prolonged vomiting and laxative abuse are the common ED causes.
- Presence of hypokalemia with a background chronic renal failure usually augurs poor prognosis.

---

* Does hypokalemia cause nephropathy? An observational study of renal function in patients with Bartter or Gitelman syndrome. Walsh, S.B. 1; Unwin, E. 2; Vargas-Poussou,

R. 3; Houillier, P. 2; Unwin, R. 1 Qjm. 104(11):939-944, November 2011.

Urine output and serum creatinine are used for classifying Acute renal failure

Glomerular Filtration Rate is used for classifying Chronic Renal Failure

Acid-base balance

- Unlike other electrolytes and minerals, the maintenance of acid—base balance is particularly strictly controlled in the body. An increase in the hydrogen ion leading to fall in blood pH is termed as acidaemia. A decrease in hydrogen ion leading to rise in blood pH is termed ad alkalaemia and the process is called acidosis and alkalosis respectively.
- The main work in maintaining acid—base balance is excreting the daily load of 70 mmol of acid contained in the diet. This is not exclusively linked to quantity of the food so patients who are significantly restricting diet might also be exposed somewhat similar amount of acid necessitating all the mechanisms to function properly to maintain the acid—base balance.
- Body primarily uses bicarbonate to get rid of the excess hydrogen ion coming into the body. Anything to upset this will result in the acid—base imbalance.
- Two organs in the body that are mainly involved in the acid—base balance are : Kidneys and lungs. Liver also plays a role in the acid—base regulation but causes more imbalance at times by using more bicarbonates (therefore increasing the rate of ureagenesis) during metabolic acidosis. In metabolic acidosis, the level of the acid is already high hence the kidneys and lungs need more bicarbonates to excrete the excess hydrogen ions.
- Two principal buffer systems are available in case of necessity for excess acid removal.

    - Titratable acid like phosphoric acid attaches the hydrogen ion to its conjugate anions and the hydrogen ion is eventually excreted through the

kidney. There is a limit to the buffering effect of titratable acids
- The ammonia buffer system can increase its capacity to carry excess hydrogen ions by hundred fold. Glutamine is an amino acid which is essential in the ammonia production to which hydrogen ion gets attached to form ammonium. This is excreted as ammonium chloride. Patient who restrict protein in their diet can in theory have a low amount of glutamine leading to limitations in this buffering mechanism.

Some numbers to remember!

- Blood pH is maintained between 7.38-7.42
- Normal adult diet contains 70-100 mmol of acid.
- Plasma bicarbonate is maintained between 22-26 mmol/L [22-26mEq/L]
- About 4500 mol of bicarbonate is filtered by kidneys every day, and almost all of this is reabsorbed.
- 4500 nmol of hydrogen ions secreted into proximal convoluted tubule.
- Liver consumes about 1000 mmol of bicarbonates every day in the process of synthesis of urea.
- Normal anion gap is 10-18 mmol/L. If we take albumin (constitute the largest part of the unmeasured anion) gap comes down to 6-12 mmol/l.
- Reduction in albumin by 1 gram will reduce the net negative charge (thereby increasing the anion gap) by 0.2-0.28 mmol/L.
- Hydrogen ion level in arterial blood gas: 35-45 nmol/L

[For all of the above, 1 mmol/L = 1 mEq/L]

Acid-base imbalance could be due to

- Impaired or excess CO_2 removal in the lungs resulting in respiratory acidosis and respiratory alkalosis respectively.
- Abnormalities in the regulation of bicarbonate and other buffer mechanisms in the blood resulting in metabolic acidosis and alkalosis.
- A study indicates progressive enlargement of peripheral lung units without relevant alveolar septa destruction in patients with anorexia nervosa. In the first 3 years of disease, appreciable weakness of respiratory muscles develops in patients with stable AN without further impairment over time. These could lead to acid-base imbalance especially if there are other risk factors*.
- Both respiratory and metabolic causes could coexist in patients who have both respiratory disease and metabolic disturbance due to renal conditions.
- On the other hand, metabolic causes could activate respiratory mechanisms e.g metabolic acidosis causes hyperventilation leading to improved removal of CO_2. This will result in partial improvement in the acidaemia, caused by the metabolic acidosis.

---

* Respiratory Function in Patients With Stable Anorexia Nervosa. Gardenghi, Giovanni Gardini MD; Boni, Enrico MD; Todisco, Patrizia MD; Manara, Fausto MD; Borghesi, Andrea MD; Tantucci, Claudio MD Chest. 136(5):1356-1363, November 2009.

Respiratory Acidosis

- In patients with extremely low weight, the respiratory muscle start to lose their power. This will result in ineffective expiration of CO_2 leading to its retention.
- Kidneys tries to compensate by increasing bicarbonate retention, but in acute respiratory compromise this will not be very effective. If the weight loss is slow, the renal retention of bicarbonate will have time to increase. Simultaneously, the concentration of hydrogen ion would have had time to return to normal. This usually takes about 2-5 days.
- Positive correlation between metabolic acidosis, BMI and percentage of weight loss are observed in a group of adolescent patients who were admitted for medical stabilisation. Bed rest and refeeding improved the condition*.

Respiratory Alkalosis

- Reduction in CO_2 leads to respiratory alkalosis. Here, the hydrogen ion concentration falls, and there will be a small reduction in the bicarbonate concentration too.
- This can be due to hyperventilation. In theory patients who exercise a lot are prone to develop this condition but the defence mechanisms set in hence this condition is not normally found in ED patients.

---

* Respiratory acidosis in adolescents with anorexia nervosa hospitalised for medical stabilisation: a retrospective study. Kerem, Nogah et al International Journal of Eating Disorders, 45(1): 125-130, January, 2012.

Metabolic causes of Acid-Base Imbalance

- Kidneys play a crucial role in the regulation of acid-base balance in the blood. The aim is to remove the excess acid (hydrogen ion) comes into the body via food and to use a base (bicarbonate) to achieve this and at the same time retaining the base.
- Two mechanisms in operation:
- Reabsorption of Bicarbonates (retaining base):

$$H^+ + HCO_3^- \leftrightarrow H_2CO_3 \leftrightarrow CO_2 + H_2O$$

A fall in the HCO_3 leads to increase in the concentration of hydrogen ions leading to Acidosis. In the body, this usually leads to increase in ventilation resulting in blunting of acidotic change in the blood. Hence the metabolic and respiratory mechanisms work in unison to maintain the balance. The above mentioned process will not result in the regeneration of HCO_3. The regeneration and maintenance of plasma HCO_3 is mainly by reabsorption of bicarbonate in the proximal convoluted tubule of the kidney.

- Excretion of Hydrogen ions (elimination of the acid) is another important process in the body. Secretion of hydrogen ions in the distal convoluted tubule of nephrons is the main source of excretion of acids. The hydrogen ions secreted in the proximal convoluted tubules are typically used to reabsorb bicarbonate ions.

- Most dietary hydrogen ions are derived from sulphur containing amino acids. Hence patients who restrict protein are more prone to develop alkalosis. The overall picture, as we know is far more complex.

Interaction with other electrolytes and elements:

- Hyperkalemia often is associated with acidosis whereas Hypokalemia is associated with alkalosis.
- High calcium level can lead to metabolic alkalosis.

Metabolic Acidosis

- This is due to accumulation of any acid other than H_2CO_3
- Number of conditions lead to metabolic acidosis e.g acid administration, acid generation-lactic acidosis during cardiac arrest, impaired acid excretion by the kidneys and more importantly for eating disorder patients increased bicarbonate loss from the gastrointestinal tract od through kidneys.
- Anion gap is useful in differentiating these possible causes

    - acidosis with normal anion gap : either HCl is retained or $NaHCO_3$ is lost. Eating disorder patients are likely to experience increased loss of bicarbonate due to laxative abuse. But the GI tract often adjusts to chronic laxative use hence this is not often seen.
    - acidosis with high anion gap : high amount of unmeasured anion like lactate or an exogenous acid gives this picture. These are usually found in plasma in small quantities. In Ed conditions, acute starvation with an resultant ketoacidosis results in this clinical picture.

- Anion gap = {[Na+] + [K+]} - {[HCO3-] + [Cl-]}
- Treatment of the underlying cause is essential.
- Severe acidosis (H) > 100 nmol/l, pH < 7.0) is associated with significant mortality.
- To correct acidosis of cardiac arrest, sodium bicarbonate (50 mol as 50 mL of 8.4% sodium bicarbonate IV is often used.

Treatment of Acidosis

[H+] > 100 nmol/L pH <7.0	Use of bicarbonate at this stage is recommended by some but this is controversial. Beware of rapid correction as this leads to fits, tetany and pulmonary oedema. Mindful of the need to improve the ventilation.
↓	↓
[H+] > 126 nmol/L pH < 6.9	Sodium bicarbonate (150 mmol/L) IV over 2-3 hours

MEDICAL WARD

Metabolic Alkalosis

- Vomiting and other types of purging behaviours in ED patients lead to metabolic alkalosis.
- Purging behaviours often result in a state of profound dehydration and chloride depletion that leads to the metabolic abnormalities. In the eating disorder patients, these abnormalities lead to a propensity towards marked edema formation resulting in PseudoBarrett's syndrome*.
- Significant mortality is associated with this condition. A pH of 7.55 is associated with 45% mortality and a pH of 7.65 lead to 80% mortality.
- Vomiting causes loss of chloride. Some diuretics like frusemide also results in loss of chloride.
- Most importantly this condition is associated with refeeding syndrome.
- Low albumin level also causes metabolic alkalosis.
- Laxative abuse leads to loss of potassium which in turn will lead to metabolic alkalosis.

---

* Pseudobarrett's syndrome in eating disorders Bahia, Amit et al : International Journal of Eating Disorders. 45(1): 150-153, January, 2012.

Metabolic Alkalosis due to Chloride depletion

- Assessment of potassium depletion is still necessary.
- GFR should be measured. If GFR is normal, bicarbonate will be excreted with sodium and potassium resulting in increased level of chloride. Hence the treatment should be aimed towards this.
- In case of extracellular volume (ECV) depletion isotonic saline could be used. If ECV is increased, hydrochloride or ammonium chloride could be used.
- If GFR is normal and alkalosis is not improving acetazolamide could be used.
- Dialysis is required if renal functions are impaired.

Metabolic Alkalosis due to Potassium depletion

- Correction of hypokalemia.
- Beware of this complication in hypokalemia. Metabolic alkalosis should be assessed in patients with significant hypokalemia.

Body's response to restriction of diet

When diet is low in protein but has enough calories:

- Excretion of urea decreases
- Uric acid excretion falls by 50%
- Creatinine excretion is unaffected
- Creatinine and 50 % of uric acid are the products of normal wear and tear hence no reduction is observed with dietary protein restriction.
- Total nitrogen excretion will not fall below 3.6 g/ day.

When diet is low in both protein and calories:

- Total nitrogen excretion raises to 10 g/day on average mainly because of use of protein for energy.
- Glucose and fat will counteract this destruction of protein.
- Most of the protein burned during starvation comes from liver, spleen and muscles. Very little comes from the heart and brain.
- An average 70 Kg man has 0.1 Kg of glycogen (storage form of Glucose) in liver, 0.4 Kg in muscles and 12 Kg of fat. Liver and muscle glycogen are enough for one day of energy requirement. This would be used very quickly after the start of starvation.

Liver Function Tests

Introduction

Liver weighs about 1.2-1.5 Kg. It is divided into right and left lobes by the middle hepatic vein. The lobes are further divided into eight segments by right, middle and left hepatic veins. Each segment has separate blood supply which enables the resection of individual segments.

About 25% of the total cardiac output reaches liver. The hepatic artery supplies 25% of the total blood flow whereas the portal vein from the gastrointestinal tract and spleen constitutes the 75%.

The acinus is the functional unit of the liver. Each acinus made up of parenchyma with portal vein radicles, hepatic arterioles and bile ductules. Hepatocytes lie adjacent to the acinus at varying distance.

Hepatocytes contains numerous mitochondria, and large amounts of rough endoplasmic reticulum and free ribosomes. They also contain cytoplasmic glycogen and lipid stores removed during histological preparation. The average life span of the hepatocyte is 5 months; they are able to regenerate. It is said that the liver can regenerate to its full size even after the removal of more than 85% of its original size.

Liver is involved in the synthesis of the protein (albumin, fibrinogen and certain clotting factors), metabolism of carbohydrates (synthesises fatty acids from carbohydrates and gluconeogenesis—synthesis of glucose, a carbohydrate from alanine, glycerol etc.) and lipids (synthesises cholesterol and bile salts) and detoxification of toxic chemicals e.g drugs and steroids produced in the body. Liver also stores large amounts of vitamins especially A, D and B12. Other vitamins like K and folate are also stored to a minor extent as well as minerals like iron, and copper.

The Biliary System

It is a complex system of bile canaliculi, ductules and bile ducts. The common bile duct receives bile juice from all the minor ductules and ducts open. Gall bladder does not take part in the synthesis of bile but stores and concentrates the bile.

Functions of the Liver

Protein Metabolism

- Liver is the principal site of production of circulating proteins apart from gamma globulins.
- Regulation of protein in the plasma is done by Liver. Plasma contains 60-80 g/L of protein which includes albumin, globulin and fibrinogen.
- Albumin is a main product of liver which is often measured to determine the functional status of the liver. About 10-12 grams of albumin is synthesised every day.
- Normal Serum range:

    Total Protein: 60-80 g/L [6-8 g/dL]

    Albumin: 31-43 g/L [3.1-4.3 g/dL]

    Globulin: 26-41 g/L [2.6-4.1 g/dL]

- Reduced albumin synthesis is often the cause for oedema in eating disorder patients.
- Liver also produces many other essential proteins like clotting factors and carrier proteins. The deficiency of these products results in certain manifestations.
- Liver also plays a role in the degradation of amino acids which is then converted to urea and excreted through kidneys.

Functions of the Liver

Carbohydrate Metabolism

- Maintenance of blood sugar is one of the major functions of the liver.
- Glycogen (storage form of glucose) in the liver is the first to be released as the first response to starvation.
- In patients with anorexia with restriction of diet, ketone bodies and fatty acids (both derivatives of fat) become the alternate source as glycogen store is rapidly exhausted.

Functions of the Liver

Lipid Metabolism

- Fats are not soluble in water which is the main base for plasma. Hence the fats are transported in the plasma as protein-lipid complexes (lipoproteins).
- Liver synthesises different forms of lipoproteins (high density lipoproteins (HDL), very low density lipoproteins (VLDL), low density lipoproteins (from intermediate lipoproteins by a liver enzyme called lipase). Other fat molecules like triglycerides, free fatty acids, cholesterol (main source is diet) are also synthesised in liver.
- There are a number complex interactions between protein, lipid and carbohydrate metabolism co-ordinated by liver. This is beyond the scope of this book.

Other functions of the liver include hormone and drug inactivation and immunological functions will not be discussed in this book. Please refer to suitable resources.

Liver Function Test Values

Alanine Aminotransferase (ALT)	3-48 U/L [3-48 units/L]
Aspartate Tramsaminase (AST)	0-55 U/L [0-55 units/L]
Total Bilirubin	upto 17 µmol/L [upto 1.0 mg/dl]
Gamma Glutamyl Transpeptidase (GGT)	8-50 units/L [8-50 units/L]
Serum Albumin	31-43 g/L [3.1-4.3 g/dL]
Alkaline Phosphatase	13-39 U/l [1.8-5.4 units/dl]

76% of patients who were admitted for medical stabilisation in an eating disorders unit with anorexia nervosa (mean BMI of around 13) showed an abnormality in the liver function test*. Isolated and mild abnormalities of liver function are common in AN patients, but their clinical significance is not very clear. Whereas, more than one abnormality or moderate to severe elevation of even one liver enzyme could have significant clinical implications. This often requires close monitoring.

---

* Severe anorexia nervosa: Outcomes from a medical stabilization unit. Gaudiani, Jennifer L. MD ; Sabel, Allison Lee MD, PhD; Mascolo, Margherita MD; Mehler, Philip S. MD, *International Journal Of Eating Disorders. 45(1):85-92, January 2012.

Liver Function Tests

Serum Albumin

- Haemoglobin has a half-life of 16-24 days and about 10-12 grams synthesised daily.
- Serum albumin measurement is useful to assess the severity of long standing liver disease.
- In acute liver dysfunction, serum albumin level may be normal.
- In eating disorder patients with significant protein restriction, early fall in albumin level is observed.
- Measurement of total protein is not particularly effective.
- Albumin plays a pivotal role in acid-base balance.

Prothrombin Time

- Prothrombin times (12-16 seconds) vary depending upon assay method hence International Normalised Ratio (INR) is used. Prothrombin time is a significant indicator for both acute and chronic liver disease.
- Vitamin K deficiency can also cause prolongation of PT hence this cause should be ruled out by Vitamin K administration (Vitamin K 10 mg intravenously).
- A study* suggests that decreased thrombopoietin production and accompanying liver dysfunction may be related to thrombocytopenia and myelosuppression in AN with severe malnutrition.

* Thrombopoietin and Thrombocytopenia in Anorexia Nervosa with Severe Liver Dysfunction. Yoshiuchi, Kazuhiro MD, PhD 1, *; Takimoto, Yoshiyuki MD, PhD 1; Moriya, Junko MD, PhD 1; Inada, Shuji MD 1; Akabayashi, Akira MD, PhD 1 International Journal Of Eating Disorders. 43(7):675-677, November 1, 2010.

Liver Function Tests

Bilirubin

- Increased serum bilirubin is usually accompanied by other abnormalities in the liver function tests.
- Determination of whether the bilirubin is conjugated or unconjugated is not usually necessary in ED patients unless haemolysis of some congenital disorder need to be ruled out.

Aminotransferases

These enzymes are located in the hepatocytes. The enzymes will leak into the blood with the damage to hepatocytes.

Two types:

- Aspartate aminotransferase: Not specific to injury to liver as the enzyme is found in heart, kidney, muscle and brain. ED patients who indulge in excessive exercise leading to muscle injury could have elevated levels.
- Alanine aminotransferase: Elevation of this enzyme is specific to liver injury.

Liver Function Test

Alkaline phosphatase (ALP)

- Present in sinusoidal and canalicular membranes of the liver.
- Alkaline phosphatase is not particularly specific to liver pathology. Widely seen in bone, intestine and many other tissues.
- If other liver enzymes e.g gamma GT is elevated, then the hepatic origin could be reasonably presumed.
- ALP can rise several fold (sometimes over 1000 IU/L) in certain conditions e.g primary biliary cirrhosis.

Gamma glutamyl transpeptidase (GGT)

- Present in microsomes of hepatocytes.
- Non specific as found in many tissues.
- GGT level is positively correlated with protein content in the diet and inversely related to fat content of young women with anorexia nervosa*.
- Alcohol can elevate the level.
- If ALP is normal, elevated GGT is a good guide to alcohol intake. Specificity of elevated GGT to alcohol intake comes down if other liver enzymes are also abnormal.

* Prevalence and Predictors of Abnormal Liver Enzymes in Young Women with anorexia Nervosa. Fong, Hiu-fai BS; DiVasta, Amy D. MD, MMSc ; DiFabio, Diane DTR, CDT c; Ringelheim, Julie BA ; Jonas, Maureen M. MD; Gordon, Catherine M. MD, MSc Journal of Pediatrics. 153(2):247-253,

Serum Amylase

Normally secreted in pancreas, small bowel and urogenital epithelium. Salivary glands secrete an isoenzyme of amylase. Difference could be made by measuring other pancreatic enzymes (e.g lipase).

Normal level: 20-110 U/L [20-110 units/L]

Endocrinology

Introduction

Endocrinology is the branch of medicine that deals with endocrine glands and hormones. Hormones are essential in a number of vital body functions and chemical processes. Both the states of deficiency and excess lead to symptoms and syndromes. In eating disorders, special status is given to endocrinology by the incorporation of loss of menstruation as one of the diagnostic criteria for anorexia nervosa. Patient also attempts to alter the metabolism by consuming thyroid hormone to alter their weight. Most important and a rather difficult combination to manage are eating disorder and diabetes mellitus. In this chapter, we will at look the relevant topics of endocrinology in the field of eating disorders.

Prior to this we will look at some of the common characteristics of hormones and endocrine glands. Hormones are chemical messengers, and they are synthesised in the endocrine glands. These messengers act on a site away from the site of production (endocrine activity), but at times these could act locally (paracrine activities). The main endocrine glands are pituitary, thyroid, parathyroid, pancreas and reproductive organs—ovary and testis. The secretion of hormone is not confined to these organs. For an example hypothalamus, part of the brain also secretes hormones as well as certain cells in the gastrointestinal tracts. The secretion by these organs may function like a hormone or may act as a neurotransmitter or a neuromodulator.

Hormone synthesis, storage and release are usually lengthy procedures requiring varying degrees of time. Whereas once released hormones have a wide range of time to create a response in the target organs. For example, insulin acts very fastly and brings down the glucose level in the blood almost immediately. Growth hormone needs time and sustenance to bring about the changes resulting in the actual growth or other changes in the individual. Thyroid hormones, used by some patients wanting to lose weight, is relatively fast acting (within six hours of consumption) and the effect lasts for days.

There are a number of features that are unique to secretion of hormones. They are:

- Most of the hormones are controlled by some form of feedback. The feedback may be positive or negative. An example is the hypothalamic-pituitary-thyroid axis.
- Hormones follow a pattern of secretion. Some of them are secreted almost continuously e.g thyroid, whereas others are phasic e.g Luteinising Hormone and Follicular Stimulating Hormone (LH and FSH). Secretion of LH and FSH is pulsatile with major pulses released once in every 1-2 hours. The secretion of these hormones in turn is linked to the phase of the menstrual cycle. The other rhythmic secretion is 'circadian' meaning secretion changes over 24 hours e.g pituitary-adrenal axis, secretion of melatonin linked to dark-light cycle of the day and night.
- Number of general factors like stress, sleep and feeding or fasting can affect the secretion of these hormones.
- The hormonal response to undernutrition is heterogeneous. In clinical practice, metanephrines, GH, and/or cortisol data could be used as valuable predictors for severe short-term outcome*.

* Hormonal Profile Heterogeneity and Short-Term Physical Risk in Restrictive Anorexia Nervosa. Estour, Bruno; Germain, Natacha; Diconne, Eric; Frere, Delphine; Cottet-Emard, Jean-Marie; Carrot, Guy; Lang, Francois; Galusca, Bogdan: Journal of Clinical Endocrinology & Metabolism. 95(5):2203-2210, May 2010.

Thyroid

- Some ED patients take thyroxine tablets to control their weight.
- T3 and T4 acts by increasing the oxygen consumption of all metabolically active tissues. They also mobilise fatty acids and increase membrane bound ATPase activity.
- The increase in the metabolic rate observed after several hours of intake and lasts at least 6 days.
- When the metabolic rate is increased rate of vitamin usage is also increased leading vitamin deficiency. Hence this group of patients are more vulnerable for vitamin deficiency.
- More potassium is liberated during protein catabolism leading to hypokalemia.
- Weight loss also occurs due to increased body heat production.
- Muscle weakness in ED patients with abuse of thyroxine could be due to thyroid myopathy.
- On cessation of intake of thyroid supplement: Look for rebound hypothyroidism and failure of TSH to increase to stimulate normal T3 and T4 secretion.

Thyroid function test

Thyroid Stimulating Hormone (TSH)

0.3-3.5 mIU/L [0.3-3.5 mIU/L]

Free T4   10-25 pmol/L [0.8-1.8 ng/L]

Free T3   3.5-7.5 pmol/L [2.3-4.2 pg/mL]

Which thyroid function test?

TSH is the single most sensitive test.

But usually a combination is needed

- TSH and free T3 or free T4 to confirm hyperthyroidism
- TSH and free T4 to confirm hypothyroidism

TRH assay is rarely tested to differentiate TSH resistance and TSHoma when raised free T4 and TSH are noted. In TSHoma, increased TSH secretion will not respond to TRH.

Sick euthyroid syndrome:

- When diet is restricted, body adjusts to it by reducing the metabolic rate. This is achieved usually by reducing the active forms of thyroid hormones and increasing the inactive form (reverse T3).
- TSH is secreted when the thyroid hormone level in the circulation is low. In the state of starvation, presence of inactive form of the hormone keeps the TSH level in check thereby stopping increased production of thyroid hormone by elevated TSH level.
- This condition should not be treated with thyroid supplement.

Can Thyroid hormone cause weight loss?

The answer is yes due to the following reasons:

- Increased heart rate and cardiac output leading to increased demand for calories by heart.
- Increased bone turnover
- Increased gut motility hence reducing the chance for food absorption
- Increased muscle protein turnover
- Increased glycolysis and lipolysis

In reality human body has various mechanisms to counteract these effects of the externally administered thyroxine. But these mechanisms can fail to lead to a clinical picture of hyperthyroidism.

Hypothyroidism

- Cessation of exogenous thyroxine abruptly can lead to hypothyroidism. High doses of thyroxine for a prolonged period can cause failure of thyroid function.
- Clinical features of hypothyroidism are as following:

    - non-specific tiredness in mild cases
    - dry hair, thick skin (due to mucopolysaccharide deposition), cold intolerance, constipation and bradycardia are seen in severe cases.
    - Weight gain can occur especially if the patient has improved the diet.

- Investigations: Please refer to appropriate pages of this book. Other abnormalities like normochromic, normocytic anaemia are seen. Rarely macrocytic anaemia is observed since associated pernicious anaemia is common in hypothyroid state.
- Increased levels of AST, Creatinine kinase, cholesterol and triglycerides are seen. Hyponatremia occurs due to increase in ADH secretion.
- Myxoedema coma is very rare.

Treatment of Hypothyroidism

- Patient should be advised against self administration of thyroxine. Information should be provided about the expected weight gain and how this will normalise once the thyroid replacement has rectified the hypothyroid status.
- Starting dose: usually 50 micrograms but in frail patient 25 microgram will be a good starting dose. ECG monitoring may be necessary especially when there is concurrent electrolyte abnormality.
- Aim to normalise the T4 and TSH levels. An increase in the dose needs at least six weeks to impact the blood levels of these hormones.
- Clinical improvement is usually evident in two weeks. Full symptom resolution may take up to six months.
- Complete suppression of TSH is dangerous especially in patients with amenorrhoea due to the combined effect of hypothyroidism and suppressed TSH leading to osteoporosis.
- Usual maintenance dose in 100-150 micrograms/day.
- During pregnancy extra 25-50 micrograms may need to be added to the maintenance dose.

Dr. Murali Sekar, Dr. Krishnakumar Muthu

Thyrotoxicosis factitia

- Self administration of thyroid supplement can cause symptoms of thyrotoxicosis (hyperthyroidism).
- As T4 is the commonly available thyroid supplement, the blood tests reveal raised T4 but not T3. This picture could arouse suspicion, but this is not confirmatory since T4 is always raised in all the other causes of hyperthyroidism and T3 is only rarely raised.
- Symptoms: restlessness, malaise, stiffness, muscle weakness, difficulty in breathing, palpitations, heat intolerance, eye signs (exophthalmos, conjunctival oedema etc).
- Symptoms like weight loss, increased appetite, vomiting and diarrhoea are part of hyperthyroidism but in ED patient group they will seldom complaint about these.
- TSH may remain suppressed for months. Hence T3/T4 levels are more reliable.
- Usually symptoms will improve within about 10-20 days.

Treatment of Thyrotoxicosis Factitia

No symptoms	Self-administration should be stopped	COMMUNITY
↓	↓	
	+	
Mild Symptoms	Propranolol 40– 80 mg; Other agents like carbimazole are seldom required	
	↓	
↓		
Symptoms of Thyroid crisis—hyperpyrexia, severe tachycardia, restlessness, cardiac failure and liver dysfunction.	Full dose Propranolol, potassium iodide, corticosteroids, supportive measures. ?Anti-thyroid drugs. Admission to ICU often necessary.	MEDICAL WARD

Diabetes mellitus (DM) and Eating Disorders

- Patients with binge eating disorder and bulimia nervosa are prone to develop type II diabetes if they are over weight.
- Patients with type I diabetes mellitus may develop anorexia nervosa or other types of ED, which will make, this group of patients one of the most difficult to manage.
- In refeeding stage of treatment for Anorexia Nervosa, an increase in abdominal fat has been reported. This tendency for abdominal deposition of fat might be associated with the onset of insulin resistance*. Hence care need to be taken to identify this as a possibility at the refeeding stage by repeated serum glucose measurement and in cases of doubt with full battery of tests e.g Glucose tolerance test.
- Insulin can often be under-dosed or omitted leading to accelerated onset of complications of diabetes mellitus. The mechanism is that the carbohydrate consumed stays in the circulation rather than being driven into cells by insulin. This process results in rapid burning off of the calories leading to loss of weight.
- In the attempt to achieve weight loss, ED patients damage the delicate insulin controlled glucose metabolism. About 200 grams of glucose is produced every day mainly from liver glycogen.
- Cognitive disturbances of varying severity are often the first and common manifestation of altered glucose metabolism. This is because brain uses almost exclusively glucose for its energy requirement, and approximately 100 grams of glucose is used by brain every day.

* In anorexia nervosa, even a small increase in abdominal fat is responsible for the appearance of insulin resistance. Prioletta, A. 1; Muscogiuri, G. 1; Sorice, G. P. 1; Lassandro, A. P. 1; Mezza, T. 1; Policola, C. 1; Salomone, E. 1; Cipolla, C. 1; Casa, Della S. 1; Pontecorvi, A. 1; Giaccari, A. 1, 2 Clinical Endocrinology. 75(2):202-206, August 2011.

Management of DM in ED patients

- Joint management by diabetic specialist and ED specialist is strongly recommended.
- Information like change of dosage of insulin and other medications, eating disorder symptoms like night eating, bingeing-purging, restriction need to be exchanged by professionals of these two specialist services.
- Patient should be advised of the impact of one condition over the other. This is achieved preferably by joint psycho-education sessions of both professional teams.

Nocturnal hypoglycaemia and Night Eating Syndrome:

Patients with the combination of insulin dependent diabetes and night eating face a particular problem. Insulin secretion falls during night time (due to not eating in the night time) but increases after 4 a.m. Night eating (bingeing) will disrupt this physiological cycle. Hence the patients can experience a high level of blood glucose due to bingeing and unavailability of insulin.

Insulin induced hypoglycaemia

- Hypoglycaemia is a common condition associated with refeeding. This is mainly due to poor glycogen reserve in this group of patients. Normally when insulin is secreted after a meal, glucagon will mobilise the stored from of glucose (glycogen) to the blood stream to maintain the blood glucose level. In refeeding stage, since the body of anorexia nervosa patients does not have much of glycogen store, this protective mechanism fails. This failure leads to hypoglycaemia.
- Hypoglycaemia is the most common complication of insulin treatment even in patients without eating disorders. This is mainly due to imbalance between diet, activity and insulin requirement.
- In eating disorders sufferers each of the components could be in disarray (e.g erratic eating pattern, excessive exercise, administration of wrong dosage, wrong timing of the dose) and a combination of all these being wrong at one time is also likely.
- Symptoms of hypoglycaemia start to occur when blood glucose falls to less than 3 mmol/L. Symptoms are usually adrenergic—sweating, tremor, palpitations. Cognitive changes are also seen.
- Beware of hypoglycaemic awareness since hypoglycaemia is common in long standing poorly controlled diabetes. This state is also common in those who experienced recurrent hypoglycaemic states.

Symptoms, Investigations and Treatment in Eating Disorders

Management of hypoglycaemia

Mild Symptoms (usually at blood glucose 2-3 mmol/L [36- 54 mg/dL] Symptoms include sweating, headache, poor concentration, palpitation ↓ Severe symptomse.g palor, drowsy, aggression, tachycardia, fits and hypoglycaemic coma (can occur at any blood glucose level < 2 mmol/L [<36 mg/dL]	Any form of rapidly absorbed glucose. Additional precautions should be taken. ↓ 50– 100 ml 50 % dextrose followed by 5% saline flush. Continuous IV infusion may become necessary rarely. If cerebral oedema + use Dexamethasone 4 mg/4h IV Rarely Glucagon 1-2 mg IM .

COMMUNITY

MEDICAL WARD

Oestrogen and Osteoporosis

- Amenorrhoea is not the cause for osteoporosis but global nutritional deficit causes both.
- When a women's body is losing weight to less than a critical point (which is usually a BMI of < 17.5), it will take necessary precautions to conserve energy and reduce the body functions which are not essential to its survival. In other words, body gives up the altruistic role of contributing towards proliferation of the human race and becomes selfish and will try to preserve its integrity and survival by retaining as much energy as possible.
- On average, a woman loses about 50-100 ml of blood during each normal menstrual cycle. When the bodily reserves are already too low, this becomes a burden which the body can do without.
- The development of osteoporosis in underweight conditions are explained in the opposite page.
- As it can be seen, any attempt to correct just one aspect of this complex mechanism will not result in the improvement of bone density e.g correction of hormone deficit without improving the mineral availability or vitamin D, calcium and phosphate supplementation without improving the protein necessary for the bone synthesis. This is the reason why 'balanced diet' is the best answer to improve this condition as it provides all that is needed.

Symptoms, Investigations and Treatment in Eating Disorders

Osteoporosis in Anorexia

Decreased food intake

↓

Weight ↓ to < a critical point

↓

Reduction in protein availability, reproductive functions become secondary

Resulting in loss of libido, sub/infertility

↓

↓ in FSH & LH

↓

↓ Oestrogen (and progesterone)

↓

↓ reduced Osteocyte synthesis

&

↑ in Osteoclastic activity (which is usually inhibited by Oestrogen)

↓

↓ Protein ——>

↓ calcium & Po4 ——>   Osteoporosis

- The pathway in the previous page is rather a simplified version as many other agents (parathormone, calcitonin, Insulin like growth factor (IGF), growth hormone, relative hypercortisolemia*) are involved in the bone synthesis and maintenance of adequate bone density.
- In men, similar mechanism causes symptoms like loss of libido, impotence and infertility.

---

* Bone health in anorexia nervosa. Misra, Madhusmita; Klibanski, Anne; Current Opinion in Endocrinology, Diabetes & Obesity. 18(6):376-382, December 2011.

Treatment of Osteoporosis due to Anorexia Nervosa

- Aim is to improve the intake of a balanced diet.
- Hormone replacement (OCP, Mini pills) is not particularly effective, but transdermal oestrogen is effective in adolescents. Recombinant preparations increases bone formation in adolescents, and with oral oestrogen increases bone mineralisation in adults with anorexia nervosa*.
- Vitamin D and calcium supplements could be tried, but the results may not be as expected.
- Insulin like growth factor (IGF-1) increases bone density by 5% after one year of treatment.
- Use of bisphosphonate need to be considered with caution due to significant side effects. Bone mineralisation shows improvement in adult patients with AN but not adolescents.
- Diagnosis and monitoring could be done with DXA scan. DXA scan exposes body to minimum radiation hence the scan is safe in that aspect. Quantitative Ultrasound of calcaneum (a bone in the foot) is an alternate choice. This method is mainly used as screening before the DXA scan.
- X-ray is usually reserved to rule out any fracture, but stress fracture may not be picked up by x-ray hence bone scan (not the bone density scan) can be used to confirm the presence of fracture.

* Bone health in anorexia nervosa. Misra, Madhusmita a, b; Klibanski, Anne, Current Opinion in Endocrinology, Diabetes & Obesity. 18(6):376-382, December 2011.

Dual Energy X-ray Absorptiometry (DXA)

- Measures areal bone density. It is also used to measure fat mass, lean tissue mass as well as bone mineral content.
- Usually lumbar spine and proximal femur are studied.
- The results of the individual patients are compared with age and gender specific norm for that individual. Result is expressed as the areal bone density of that patient in relation to the standard deviation i.e a result of -2.5 means the bone density of the scanned individual is 2.5 SDs (standard deviation) less than the expected mid-point for that individual after considering her/his age and gender.
- This method is considered as the gold standard in osteoporosis diagnosis.
- Presence of osteophytes, spinal deformity and fractures can influence the density at a particular point on the bone.

Combination of Calcium 700-100 mg/day and Vitamin D 400-800 IU is recommended.

Other rarely used medications:

Bisphosphonate e.g Alendronate 70 mg/ day

Strontium 2 grams/day

Raloxifene (selective oestrogen receptor modulator) 60 mg/day

Hormone replacement therapy (HRT)

Calcitriol, Calcitonin Insulin like Growth Factor (IGF-1)

Functional Hypopituitarism

- Anorexia nervosa with starvation can cause 'functional hypopituitarism.
- Symptoms and signs of hypopituitarism will be the same as primary deficiency of peripheral endocrine glands e.g decreased production of FH/LSH will be the same as primary ovarian failure in terms of reduction in the levels of progesterone and oestrogen.
- In functional hypopituitarism, the levels of stimulating hormones secreted by pituitary (LH/FSH, TSH, GH, ACTH) may not be markedly low. The failure is mainly due to the absence of substrates that are necessary for these hormones to act.
- An example: TSH increases the levels of thyroid hormones, which in turn, increases the rate of glycolysis/ gluconeogenesis and intestinal glucose absorption. If there is not enough glucose is consumed by the patient with anorexia, none of the above will happen to lead to 'functional' hypopituitarism.
- Interaction between GH and Ghrelin is not clear, but the levels of Ghrelin are closely related to BMI of weight stable individuals i.e higher the BMI lower the level of ghrelin. The concentration is high just before a meal and comes down in the fed-sate. Level is elevated in AN patients.
- Symptoms depend on the defect in the particular stimulating hormone-target endocrine gland axis. E.g TSH—Hypothyroidism leads to tiredness, malaise,

GH deficiency causes growth failure in children and impaired well being in adults.
- Growth Hormone levels are directly influenced by state of nutrition. In the fed state, the levels are low and during starvation level goes up.*

Treatment of functional hypopituitarism

- Improve the nutrition.
- Replacement of stimulating hormones is not necessary.
- If symptoms persist after improving the nutrition and weight, hormone assay could be done to find out which line is particularly deficient.
- Symptoms tend to improve at varying rates as nutrition improve.

---

*   Does the pituitary somatotrope plays a primary role in regulating GH output in metabolic extremes?. Luque, Raul M. 1; Gahete, Manuel D. 1, 2; Cordoba-Chacon, Jose 1; Childs, Gwen V. 3; Kineman, Rhonda D. 2 Annals of the New York Academy of Sciences. 1220(1):82-92, March 2011.

Other substances involved in eating/fasting:

- Neuropeptide Y (NP Y) and agouti-related protein (AgRP) are involved in the central appetite stimulating pathway. They have an origin in ventromedial part of arcuate nucleus. This pathway increases appetite and decreases spending of energy leading to weight gain.
- Alpha-melanocyte stimulating hormone is involved in the central appetite-suppressing activity. This is secreted from dorsolateral part of the arcuate nucleus. This hormone reduces appetite by increasing satiety and increases energy expenditure leading to weight loss.
- Ghrelin is involved in peripheral appetite-stimulating effect. It is secreted by oxyntic cells in the stomach. This acts by stimulating the appetite. Hence the level is high during starvation and low in the fed-state.
- Leptin and insulin are involved in peripheral appetite-suppressing action. Peptide Y produced in the L cells of the large intestine and distal small bowel also acts by suppressing the appetite via NPY and AgRP. Endo cannabinoids are also implicated in appetite suppressing effect.
- Agouti related protein (AgRP) found in hypothalamus increases the food intake and reduces the spending of energy during periods of starvation. Evidence so far indicates that levels of AgRP are influenced by nutritional status rather than AgRP influencing the nutritional intake leading to starvation or feeding[1].
- Another neuropeptide apelin (APE) is involved in the control of appetite and food intake. APE is secreted in the organs involved in the control of hunger and satiety: the stomach, hypothalamus, and fat tissue.

There are two principal variants namely APE 36 and APE 12. Serum APE-36 and APE-12 concentrations decreased as a result of fat tissue depletion in patients with AN. Conversely obese adolescents had elevated APE-36 and APE-12 due to excessive fat mass as well as increased APE production in adipose tissue[2]

[1.] Agouti-related protein in patients with acute and weight-restored anorexia nervosa. Merle, J. V. 1; Haas, V. 1, 2; Burghardt, R. 1; Dohler, N. 1; Schneider, N. 1; Lehmkuhl, U. 1; Ehrlich, Psychol Med. 2011 Oct;41(10):2183-92

[2.] Assessment of serum apelin levels in girls with anorexia nervosa. Ziora K, Oświecimska J, Swietochowska E, Ziora D, Ostrowska Z, Stojewska M, Klimacka-Nawrot E, Dyduch A, Błońska-Fajfrowska B J Clin Endocrinol Metab. 2010 Jun;95(6):2935-41.

Fluid Balance

Some facts about body water

- Body water constitutes 50-60% of weight in men and 45-50% in women.
- In a 70 Kg adult male, total body water is approximately 42 L.
- Three major compartments in which the body water is stored.

    - Inside the cells (Intracellular)-28 L (35% of body weight)
    - Outside the cells (extracellular)-9.4 L (12% of body weight
    - Inside the blood vessels (arteries, veins and capillaries)-4.6 L (4-5%)

- Small amounts are seen in bone, connective tissue, digestive tract and cerebrospinal fluid.
- All these three major water compartments have one key solute (water being the solvent) dissolved in it. Intracellular fluid has potassium, extracellular has sodium and plasma has proteins as their major solute.
- Relevance of body water in eating disorders: oedema during re-feeding, low protein associated oedema, water loading prior to being weighed by professionals, significant water restriction along with food restriction, habitual water loading. Body water needs to be tightly controlled since it can affect concentration of various solutes in them leading to clinical deficiencies of sodium, potassium etc.

Dr. Murali Sekar, Dr. Krishnakumar Muthu

Extracellular fluid and the impact of various agents

Agent	Main site of action	How	Effect
Effective Arterial Blood Volume (EABV)	Glomerulus and tubules	Renal sodium and water excretion through GFR and tubular reabsorption	When EABV is expanded, urine sodium excretion increases and vice versa
Renin–angiotensin–aldosterone system	Glomerulus and distal tubule	Renal sodium and water excretion through GFR and tubular re-absorption	Increase in the activity reduces sodium and water excretion.
Atrial Natriuretic Peptide (ANP)	Collecting ducts	Through volume receptors by increasing sympathetic activity and secretion of catecholamines	Reduction in volume lead to increase in catecholamine and sympathetic activity
Anti-diuretic Hormone (ADH)	Cortical and medullary collecting ducts	Increasing the water permeability of the ducts.	Decreased reabsorption of sodium and chloride and increased reabsorption of water

Regulation of Plasma Volume

- ADH plays a central role in the maintenance of plasma osmolality. When the plasma osmolality is less than 275 mOsm/Kg (Na 135-137 mmol [135-137 mEq/L]) no ADH will be found in the circulation. As osmolality raises (sodium level raises), ADH secretion also increases producing anti-diuresis, in other words retains water.
- In ED following clinical situations could be encountered:

    - Ingestion of water load: plasma osmolality will go down initially (due to more water with the same amount of sodium)
    - no ADH will be secreted leading to diuresis resulting in the elimination of the extra water. This is usually completed in 4-6 hours. Usually this system is so effective, and the other systems like the release of Atrial natriuretic peptide (ANP) rennin-angiotensin-aldosterone are not activated.
    - On the other hand water depletion results in the secretion of ADH and stimulation of thirst resulting in water retention.

Regulation of Cell Volume

- Increase in cell volume is usually dealt with an increase in the efflux of potassium from the cell leading to regulatory volume decrease of the cell volume. Other osmolytes like chloride, taurine and amino acids are also transported out to reduce the cell volume.
- Under hypertonic conditions, (a decrease in cell volume leading to shrinkage) the movement osmolytes happens in the reverse direction i.e potassium, chloride and other osmolytes move into the cells.

Abuse of Diuretic agents

- Any form of diuretic can be abused by patients as most of these are available in the internet.
- Weight loss is not just due to water loss. Thiazide diuretics promotes protein breakdown.
- Diuretics are grossly inefficient in causing loss of weight due to the following reasons:

- After the initial reduction in the body water, GFR rate comes down due to constriction of glomerular arterioles leading to reduction in the amount of water entering the tubules and subsequent excretion.

Other mechanisms like rennin-angiotensin-aldosterone get activated. This prevents loss of water beyond a critical point hence weight loss due to water loss is stopped

- Oedema is observed with diuretic abuse as well as laxative abuse.

- Treatment is usually termination of the diuretic usage. Rebound oedema following discontinuation of laxative or the diuretic should be watched for carefully. Very rarely potassium sparing diuretics are used to treat any rebound oedema.
- Potassium sparing diuretics like spironolactone (50-200 mg) is preferred in treating oedema in the re-feeding syndrome.

Diuretic agents

Group	Example	Action	Selected Side Effects
Loop Diuretics (most potent)	Furosemide, Bumetanide	Potent in stimulating water excretion in the states of water overload.	Hypokalemia, hypomagnesemia, myalgia (bumetanide), decreased glucose intolerance
Thiazide Diuretics (potent)	Bendroflumethiazide, Metalozone, Indapamide	Blocks sodium chloride channel. May Interfere with water excretion leading to water retention.	More chance for urate retention, glucose intolerance and hypokalemia than loop diuretics.
Potassium-sparing Diuretics (less potent)	Spironolactone, Amiloride, Triamterene	Spironolactone reduces sodium absorption. Other two inhibit sodium reabsorption.	Hyperkalemia, gynaecomastia in men
Carbonic anhydrase Inhibitors	Acetazolamide	Inhibite carbonic anhydrase enzyme	Metabolic acidosis and hypokalemia
ADH receptor blockers (aquaretics)	Lixivaptan, Tolvaptan	Blocks the receptors for the vasopressin	

Miscellaneous Topics

Neutropenia

- Defined as an absolute neutrophil count of < 1.0 x 10^9/L.
- Following grading based on cell count is useful:

  1.5-1.0 x 10^9/L : No significant elevated risk of infection.

  0.5-1.0 x 10^9/L : Elevated risk: can be treated as out-patient with close

  Monitoring

  < 0.5 x 10^9/L : Major risk; Treat all fever with a broad spectrum antibiotics.

  Isolation to be considered to prevent opportunistic infection.

- Usually fever is the first sign. Beware that sometimes patients may present without pyrexia, especially if they are on steroids.
- Urgent clinical assessment, blood culture and commencement of broad spectrum antibiotics are essential.
- In severe cases, evidence for cardiovascular shock, respiratory failure, tachycardia, hypotension, peripheral vasodilation are often seen. Cardiac arrest even though rare is the usual cause for death.
- Gram (+) ve infections are more common than gram (-) ve infections.
- In anorexia, the common blood picture is normochromic anaemia ± leucopenia (WBC count < 4.0 x 10^9/L). Bone marrow is typically hypocellular.

Management when cell count is < 0.5 x 10^9/L

- Advanced life support if the patient is in cardiac arrest
- Other measures:

  Rapid infusion of albumin of albumin 4.5% to restore BP if hypotensive.
  Central catheter to monitor CVP.
  O2 if Oxygen saturation is less than 95%.
  If the platelet count is less than 20 x 10^9/L—stop bleeding by pressure over bleeding site for 30 minutes. Platelet infusion may be necessary.
  Septic screen; First dose of the first line of antibiotic should be started e.g Ureidopenicillin, aminoglycosides (ceftazidime, ciprofloxacin)
  Second line antibiotics like vancomycin may be necessary.
  Continue antibiotics at least for 7 days.

Symptoms, Investigations and Treatment in Eating Disorders

Laxative Abuse

Physiology of defecation

- Rectum and anus are normally empty.
- Two types of movements occur in the colon. Non-propagative movements help in mixing the contents of the colon. Propagative movements help the content to move towards rectum.
- When there is distension in the colon due to accumulating stools, neuroendocrine cells start to secrete serotonin. This activates the sensory neurons and the signals are transmitted to the brain. This results in the propagatory movement in the colon leading to movement of the stool into rectum. Sensation of fullness ensue in the rectum and urgency to defecate happens when the rectal contents reach a threshold of 100 mL. A co-ordinated activity of relaxation of the internal sphincter, rectal contraction, relaxation of the external sphincter and puborectalis muscle results in actual expulsion of faeces.
- Normally only 150 mL of water, 5 mmol [5 mEq/L] of sodium and 12 mmol [12 mEq/L] of potassium excreted in the stool in 24 hours. Laxatives, by their action of increasing the colonic secretion and reducing the transit time for the stool in the colon, reduces the chance for the electrolytes to be absorbed back into colonic circulation.
- Due to the loss of fluids and electrolytes in laxative abuse, the rennin-angiotensin-aldosterone system will be overactive. The result will be an increased tendency to retain fluids by the body. This is the

reason why abrupt cessation of laxatives leads to oedema. Body needs time (usually 7-14 days) to downregulate the rennin-angiotensin-aldosterone system to stop retaining fluids excessively.

Laxatives

Group	Example	Action	Selected Side Effects
Bulk-forming Laxatives	Dietary fibre Wheat bran Methylcellulose	Increases the content of the colon leading to increase in firing by nerves of colon resulting in defecation	Flatulence
Stimulant Laxatives	Bisacodyl Senna and Dantron Sodium docusate	Stimulates colonic contractility directly and causes intestinal secretion	More chance for electrolyte loss in the intestinal secretion.
Osmotic Laxatives	Magnesium Sulphate Lactulose Macrogols	Increases colonic secretion; soften the stool and stimulate colonic contractions	More chance for electrolyte loss. Flatulence (uncommon with macrogols)
Irritant Suppositories	Bisacodyl Glycerol	Stimulates colon	Can cause electrolyte loss.
Enemas	Arachis oil Hypertonic phosphate Sodium citrate		Major electrolyte loss if not used judiciously.

Laxative abuse

- Stimulant laxative abuse is the most common type of abuse.
- Patient needs to be informed that abuse of laxative is not an effective way to lose weight (for some patients this is a way to 'clean' their body). The fluids lost in the stool will be regained by increased fluid retention by the rennin-angiotensin-aldosterone system. At times, the amount of water retained may exceed the water lost in the stool. This could result in edema and increase in the weight.
- Chronic laxative abuse can lead to cathartic or atonic colon. Melanosis coli is a complication of prolonged use of chronic anthranoid type (e.g senna) of laxatives. Animal studies show malignant transformation of melanosis coli.

Treatment of laxative abuse

- Key is to aim for avoidance of constipation, electrolyte abnormality and oedema. Rate of laxative withdrawal could be decided based on the above three signs.
- Method 1: All the stimulant laxatives should be stopped*. A combination of docusate, mucilloid, and magnolax will then be commenced. The dose of these will be gradually tapered down. To watch out for those signs and the rate of withdrawal will be altered based on that. Aim for pre-morbid level of bowel movements. Laxatives, commonly, are more than just a purging agent for the patients. It has other functions like cleansing, relief from guilt feeling. Hence these factors should also be included in dealing with laxatives abuse.
- Method 2: Abrupt cessation of all the laxatives and treatment of arising complication is tried in certain specialist inpatient units. Some patients might find this too difficult.
- General guidelines:
    - Adequate water and fibre intake should be maintained.
    - Patient to be explained of the possible effects of laxative withdrawal.
    - Reassurance to be provided about the transient nature of oedema if occurs.
    - Psychological dependence i.e feeling impure without laxatives need to be addressed through therapy.

* Combat laxative abuse by stopping stimulant laxatives and establishing healthy gastrointestinal function. Drugs & Therapy Perspectives. 27(3):12-14, March 1, 2011.

Difference in metabolism and dietary needs between adults and children

- Higher intake of protein and energy is recommended for adolescents to meet the increased demand due to the faster rate of growing.
- For most micronutrients, recommended daily intake is the same as for adults except few like calcium and phosphorus.
- Children and adolescents need more water (fluids) than adults.
- Basal metabolic rate decreases as age increases. By the age of 10, a child's BMR will be 80% of its BMR when the child was one year old. The rate of reduction in the BMR lessens somewhat during puberty. By the age of 20, the BMR is only 70% of what is seen at one year of age. Energy supply and BMR need to be considered in calculating the nutritional requirement of a child or adolescent recovering from anorexia.
- 80% of the energy expenditure from BMR is by heart, liver, brain, kidneys and skeletal muscles. As the child grows the percentage of metabolic rate of individual organs also changes. There is an increase in the % of metabolic rate (proportion of the energy used by the organ of entire metabolic rate of the body) in the liver and heart, whereas the % of metabolic rate falls in the brain, kidneys and skeletal muscles.
- Starvation or restriction of food intake leads to reduction in BMR. Absolute starvation leads to reduction in BMR by 75% in about 20 days.
- Potassium, magnesium and calcium deficiency can sometimes increase the BMR.

- Children weighing less than 30 Kg, an estimate of their metabolic rate may be reasonably made from their body weight alone. Their age and gender may not be particularly relevant in the calculation of BMR.

Amphetamine Abuse (Sympathomimetics, psychostimulants, analeptics)

- This class of drugs are used as a way of losing weight.
- Mechanism of weight loss is not clear but release of dopamine in the ventral tegmental area and the resulting activation of reward circuit pathway might have a role. Other hypothesis is that the weight loss is not linked to this pathway at all.
- Other physical symptoms like tachy or bradycardia, dilated pupil, altered blood pressure, sweating, muscle weakness and agitation are suggestive of a possible amphetamine use.
- Evidence for dependence and withdrawal is often evident.
- Treatment often depends on the level of dependence. Diazepam is often used to control agitation or restlessness.
- Bupropion may have a value in controlling symptoms in the withdrawal state. Psycho-education of unwanted effects, supportive psychotherapy are particularly helpful

Creatine Kinase

An enzyme found in muscle and brain.

Three isoenzymes (BB, MM and MB)are found in the human body. Most common isoenzyme found in the serum is MM.

Increase in plasma CK activity is seen skeletal or heart muscle (following myocardial infarction or ischaemia) damage.

In ED setting elevated CK activity could be due to skeletal muscle damage following excessive exercise or in extreme state of malnourishment, the muscle membrane could lose their integrity leading to an increased CK level. This is usually a sign of grave concern.

In case of doubt about the source of increased CK, isoenzyme estimation would help us to locate the source i.e a finding of >5% of total CK is of MB isoenzyme then it is highly likely that the elevated enzyme activity is due to cardiac muscle damage.

Normal range: 20-220 IU/L [0.33-3.67 mckat/L]

5 times the upper normal limit: could be physiological (especially in African-Caribbeans), in patients who take statins, hypothermia or hypothyroid state.

5-10 times the upper normal limit: seen in severe exercise, or even in moderate exercise when the patient is of low weight.

More than 10 times the normal limit: extremely rare. Seen in rhabdomyolysis. Acute renal failure and other life threatening complications are imminent.

Treatment: In ED setting, cessation of exercise would lead to fall in the level. Treating the underlying cause or any organ failure is necessary.

Glossary of Terms and Definitions

Addison's disease: Primary hypoadrenalism; a rare condition affecting adrenal glands (an endocrine gland situated adjacent to Kidneys); commonly caused by autoimmune diseases.

Acanthosis: means increased thickness of the epidermal layer (stratum spinosum) of the skin.

Angular stomatitis: skin lesion seen in the angle of the mouth where the facial skin meets the mucosal surface of the mouth (buccal mucosa). This condition is observed in vitamin B deficiencies and iron deficiency.

Arcuate nucleus: a crescent shaped aggregation of neurons in the basal part of the hypothalamus. The nucleus has various parts like ventromedial, dorsolateral (names are based on the positions in the nucleus).

Ascites: collection of fluid in the abdominal cavity formed by peritoneum (membrane seen in the abdomen).

Bilirubin: is a breakdown product of 'heme' the pigment in the red blood cells.

Bitot's spots: accumulation of keratin (a type of protein) deposits on the superficial layer of conjunctiva: usually oval or triangular in shape but can be irregular too.

Cardiomyopathy: any condition affecting heart muscles e.g genetic conditions, diabetes mellitus, hyperthyroidism and rarely poor nutrition.

Cerebral Oedema: Swelling of the cells of the brain or interstitium due to increased permeability of blood vessels in the brain. This often results in the rise in intracranial pressure.

Cholestatic jaundice: a type of jaundice resulting from the obstruction to the flow of bile acids.

Conjunctival xerosis: dryness of conjunctiva

Diabetes insipidus: A clinical condition characterised by impaired reabsorption of water by kidneys. This is either due to reduced antidiuretic hormone (ADH) secretion by the pituitary gland or impaired response to ADH by kidneys.

Diabetic Ketoacidosis: a clinical state associated with Type I diabetes mellitus and characterised by dehydration, acidosis and coma. It is one of the diabetic emergency requiring urgent treatment.

Erythropoiesis: production of red blood cells.

Exophthalmos: Protrusion of eye ball usually due to deposition of various elements in the space behind the eye ball. Seen in hyperthyroidism, cancer in the space behind the eye ball(retro-orbital tumours) etc.

Ferritin: a type of protein that stores and releases iron when and where necessary.

Haemolytic anaemia: anaemia due to destruction of red blood cells. Causes include hereditary condition like sickle cell anaemia, autoimmune disorders, cardiac anaemia (trauma)

Hyperuricaemia: abnormally high level of uric acid in the blood (women > 6 mg/dl; men > 6.8 mg/dl).

Ileus paralyticus: a functional (not mechanical) obstruction of bowel mobility. Caused by conditions like hypokalemia, post surgery, inflammation of the peritoneum.

Keratosis: accelerated growth of keratin (a layer of skin).

Lactic acidosis: increased level of lactic acid in the blood leading to metabolic acidosis; due to conditions like diabetes, vigorous, muscular exercise.

Lipoproteins: a complex biochemical structure that contains both lipids and protein e.g many enzymes, structural proteins, antigen

Megaloblasts: red blood cells which are larger than the normal size (mean volume of the cells > 96 femtolitres).

Microcytes: red blood cells whose volume is less than 76 femtolitres.

Myoclonic jerks: sudden, involuntary twitchy or jerk like focal or generalised muscle contraction; can be benign but often caused by conditions affecting the spinal cord, brain stem or cerebral hemispheres.

Myxoedema: a specific form of oedema due to deposition of connective elements (mucopolysaccharides, hyaluronic acid etc) in the layer just beneath the skin (subcutaneous layer). Seen in hypothyroidism and Grave's disease.

Oligomenorrhoea: reduction in the menstrual flow. The reduction can be in the actual flow or number of days or both.

Osteomalacia: softening of bone due to defective bone mineralisation; usually due to low availability of calcium and/or phosphorous for bone mineralisation. Similar problem in children is known as rickets. Causes: malnutrition, malabsorption, chronic renal failure etc.

Papilloedema: swelling of optic disc which contains nerve fibres from rods and cones of the eye. Any cause for a raise in the intracranial pressure usually leads to papilloedema.

Pernicious anaemia: a type of anaemia caused by loss of parietal cells in the stomach. These factors are responsible, in part, for the secretion of Intrinsic Factor, a protein essential for subsequent absorption of vitamin B12.

Pleural effusion: collection of fluids in between the layers of pleura (membrane covering lungs).

Polyneuropathy: a neurological condition involving both motor and sensory (not always) nerve fibres. Caused by autoimmune causes, diabetes mellitus, liver or kidney failure.

Polyuria: increased urine output i.e > 3.5 L/24 hrs). Common causes include diabetes mellitus, polydipsia, hypercalcemia, diabetes insipidus etc.

Pre-renal Uraemia: an excess in the blood of urea, creatinine, and other nitrogenous end products of protein and amino acid metabolism due to various causes (e.g burns, severe dehydration resulting in the reduction of blood flow to kidneys (also called pre-renal azotaemia).

Psoriasis: common skin condition characterised by well demarcated red scaly plaques. The skin cells in the affected area multiply at 10 times the normal rate.

Psychogenic polydipsia: drinking more than 3L of water per day and the cause being a mental illness including water loading observed in ED patients.

Rhabdomyolysis: Injury or necrosis of muscle fibres; caused by trauma, excessive exercise, inflammation of muscle.

Rhodopsin and Iodopsin: pigments associated with perception of light in the dark and day light conditions, respectively.

Syncope: loss of consciousness; causes include vasovagal syncope, situational syncope e.g micturition, cough or cardiac syncope

Transferrin: a type of protein involved in carrying iron molecules in the blood.

Wernicke-Korsakoff Syndrome: Wernicke is an acute condition associated with triad of nystagmus (poor coordination of eye movements), confusion and ataxia (gait or walking abnormality). Korsakoff is a chronic condition presenting with features of dementia.

Further reading list

1. Kumar and Clark's Clinical Medicine Seventh Edition; Edited by Parveen Kumar and Michael Clark : Saunders Elsevier 2009
2. Oxford Handbook of Clinical Medicine (Oxford Medical Handbooks) by Murray Longmore, Ian Wilkinson, Edward Davidson and Alexander Foulkes OUP Oxford; 8 edition
3. Harper's Illustrated Biochemistry, 28th Edition (Lange Basic Science) by Robert K. Murray, Victor W. Rodwell, David Bender and Kathleen M. Botham McGraw-Hill Medical; 28 edition
4. Ganong's Review of Medical Physiology, 24th Edition Kim E. Barrett, Susan M. Barman, Scott Boitano, Heddwen L. Brooks ; McGraw-Hill Medical; 24 edition
5. Oxford Handbook of Nutrition and Dietetics Joan Webster-Gandy, Angela Madden and Michelle Holdsworth OUP Oxford 2006.
6. Medical Management of Eating Disorders—Second Edition: C. Laird Birmingham and Janet Treasure; Cambridge University Press 2010.
7. Eating Disorders: A guide to medical care and complications. Mehler, P.S. and Anderson, A.E. Baltimore, MD: Johns Hopkins University Press 2010.

Index

A

acetazolamide 156
acidaemia 147, 149
acidosis 46, 48, 82, 147, 149, 150, 152, 153, 226, 227
Acrodermatitis 69, 70, 111
acrodermatitis enteropathica 69, 70
ACTH 198
acute bronchitis 16
Addisons disease 42, 225
agouti-related protein 200
Alanine Aminotransferase 166, 169
alcohol 24, 28, 69, 88, 97, 106, 170
alcohol consumption 88, 176, 178
alcohol dehydrogenase 69
aldosterone 46, 47, 50, 135, 207, 208, 215, 218
Aldosterone 47
alkalaemia 147
alkaline phosphatase 69, 170, 126, 166
alkalosis 46, 48, 66, 147, 149, 150, 152, 155, 156, 157
Alpha melanocyte stimulating hormone 200
Alzheimers disease 128
amenorrhoea 34, 145, 183, 192
ammonia 141, 148
Amphetamine 222

anaemia 4, 5, 20, 21, 25, 26, 79, 80, 108, 115, 182, 213, 226, 228
Angiotensin 47
angular stomatitis 103, 106
antidiuretic hormone 135, 226
aortic stenosis 17
apelin 200, 201
aplastic anaemia 9
appendicitis 31
arrhythmia 16, 17, 54
Aspartate aminotransferase 169
Aspartate Tramsaminase 166
ataxia 25, 97, 98, 123, 130, 229
atrial fibrillation 21
atrial or ventricular ectopics 21
atrophic glossitis 7
autonomic neuropathy 22

B

B12 deficiency 21
Beaus lines 4
beef 69, 104
benzodiazepines 24
bicarbonate 50, 135, 147, 148, 149, 150, 151, 152, 153, 156
bicarbonates 147, 148
bile 31, 76, 113, 131, 161, 162, 226
bilirubin 31, 32, 168, 225
Biotin 88, 110, 111
bisphosphonate 66, 195, 196
bitots spots 121
Blood urea nitrogen 141

231

bradycardia 16, 18, 20, 21, 182, 222
Bradypnoea 23
broccoli 58, 95, 113
bronchiectasis 4, 22
bronchospasm 16
brown rice 95
Bupropion 222

C

caeruloplasmin 73
calcitonin 194
calcitonin 60, 197
calcitriol 60, 197
calcium 27, 37, 46, 58, 59, 60, 61, 63, 65, 66, 68, 124, 126, 152, 193, 195, 220, 228
calcium 38, 50, 58, 60, 127, 196
candidiasis 6
cardiomyopathy 83
carotinaemia 31
cereals 58, 73, 77, 102, 108, 124, 128
cheilosis 103
chest pain 10, 16, 48
chicken 104
chromium 38
chronic laryngitis 18
chronic renal failure 5, 145, 228
Clotting factor 131
clubbing 4
cobalamin 93
cobalt 38
coma 27, 41, 82, 98, 182, 226
confusion 23, 25, 26, 44, 229
conjunctival oedema 184
convulsions 26
COPD 4, 16
copper 38, 72, 73, 76, 162
Creatine 139, 140, 222
creatinine 105, 130, 136, 137, 138, 139, 140, 143, 146, 157, 182, 228
creatinuria 139, 143
Cyanocobalamin 93
Cyanosis 17

D

dairy foods 40
death 10, 41, 213
dementia 106, 229
dermatitis 70, 106, 110, 111
desferrioxamine 82
diabetes 44, 52, 54, 175, 186, 189, 190, 225, 226, 227, 228
diabetes insipidus 44, 228
Diabetes Mellitus;DM 186
diabetic amyotrophy 28
diabetic ketoacidosis 66
diarrhoea 41, 46, 47, 54, 76, 82, 106, 113, 184
diuretics 17, 41, 42, 46, 47, 54, 63, 66, 155, 208, 209
dizziness 16, 22, 23, 25, 26, 41
Dry Beri-Beri 99
Dyspnoea 16
dysuria 25

E

Ebsteins anomaly 18
ECG 16, 20, 21, 27, 50, 183
eggs 40, 69, 83, 91, 95, 102, 104, 110, 112, 124
exophthalmos 184

F

ferritin 4, 77, 79, 80, 226
Ferrous sulphate 82
Fish 40, 46, 58, 91, 119, 124
folate 21, 91, 95, 115, 162
Folic acid 5, 87, 92, 95
Free T4 18, 179
fruits 31, 46, 95, 113, 119
FSH
   Follicle stimulating hormone 34, 176, 198

G

Gamma glutamyl transpeptidase 166, 170
gamma GT 170
Gastric carcinoma 31
gastroenteritis 31
GGT 166, 170
GH
   Growth Hormone 111, 176, 198, 199
Ghrelin 198, 200
glossitis 106, 110
Glucagon 17, 26
growth hormone 194
gynaecomastia 7
Gynaecomastia 7

H

Haematemesis 22, 32
haemoglobin 4, 9, 77, 79, 80, 108, 167
haemoglobinopathies 18
Haemoptysis 22
haptocorrin 93
headache 23, 97, 123, 130
heart disease 4, 52, 58, 74
Heberdens nodes 5
hepatitis 31
Hoarseness 18
Hormone replacement therapy 197
hypercalcemia 56, 63, 64, 124, 228
hypercholesterolaemia 105, 110
hyperglycaemia 17
hyperkalemia 23, 50
hyperkalemia 50, 51, 145, 152
hypermagnesaemia 56
hypermagnesaemia 56
Hypernatraemia 44, 45
hypernatremia 44
hyperparathyroidism 66
hyperphosphataemia 68
hypertension 21, 52
hyperthyroidism 4, 179, 181, 184, 225, 226
Hypoalbuminaemia 5
hypocalcemia 23, 54, 61
hypocapnia 23
hypoglycaemia 25, 27, 28, 112, 189, 190, 191
Hypoglycaemia 17, 25, 26, 190
Hypokalemia 9, 16, 20, 23, 47, 54, 145, 157, 178, 227
Hypokalemia 21, 47, 49, 91, 152
Hypomagnesaemia 48, 54, 55
hyponatraemia 27, 41, 42, 43, 182
hypophosphataemia 66
hypopituitarism 198, 199

hypotension 16, 17, 21, 22, 25, 27, 82, 213
hypotension 21, 27
hypothermia 6, 20, 24, 223
Hypothermia 6, 20
hypothyroidism 9, 18, 42, 178, 179, 182, 183, 227
Hypothyroidism 20, 25, 182, 183, 198
hypoxia 23, 25, 77

I

IGF-1 195, 197
ileus paralyticus 50
immunoglobulins 9
immunosuppressants 9
insulin 17, 46, 48, 175, 186, 187, 188, 189, 190, 200
Insulin like Growth Factor 197
Insulin-purging 17
intrinsic factor 91, 93
iodine 38
iron 4, 5, 9, 33, 38, 39, 72, 77, 78, 79, 80, 81, 82, 91, 102, 113, 115 162, 225, 226, 229
ischaemic heart disease 4, 128
ischemic heart disease 21

J

Jaundice 31, 131

K

Keratomalacia 121
kidney 56, 60, 95, 113, 135, 145, 148, 151, 169, 228
Koilonychia 4
Kyphosis 24

L

Lamb 69
Lanugo hair 5, 7
laryngeal carcinoma 18
laxative abuse 42, 121, 132, 145, 152, 208, 215, 218, 219
Laxative Abuse 215
laxatives 33, 56, 91, 110, 216, 218, 219
legumes 46, 52, 73
Leptin 200
lethargy 25, 41
leucopenia 213
LH
   Luteinising hormone 34, 176, 198
liver 4, 8, 9, 28, 32, 41, 42, 47, 66, 76, 88, 95, 102, 105, 110, 119, 113, 123, 124, 125, 139, 140, 141, 142, 147, 148, 157, 159, 161, 163, 164, 165, 166, 167, 168, 169, 170, 186, 220, 228
lung abscess 4, 22

M

magnesium 27, 38, 46, 52, 53, 54, 56, 61, 98, 220
maize 104
Mallory Weiss tear 32
manganese 38
meat 40, 46, 65, 77, 83, 91, 108, 112
Mees lines 5
megaloblastic anaemia 5

melaena 33
Melanosis coli 218
Menkes kinky hair syndrome 73
metabolic acidosis 147, 149, 227
Metabolic alkalosis 155, 156, 157
milk 58, 69, 73, 91, 95, 97, 102, 104, 110, 119, 124, 131
molybdenum 38
multiple sclerosis 22
myocardial infarction 20, 222
myocardial ischaemia 20
myoglobin 77
myoglobinuria 143
myopathy 178
Myxoedema 182, 227

N

neuromyopathy 28
Neuropeptide Y 200
Neutropenia 213
neutrophil 213
niacin 88, 104, 106, 108
niacinamide 88
nicotinic acid 7
Night blindness 70, 121
nuts 52, 58, 73, 97, 128
nystagmus 25, 98, 229

O

OCP 8, 195
Oesophageal carcinoma 30
Oesophageal stricture 30
oestrogen 7, 8, 195, 192, 197, 198
oligomenorrhoea 145
Onycholysis 4
Onychomedesis 5
ophthalmoplegia 25
Orthopnoea 16
osteomalacia 28, 126
osteoporosis 8, 21, 24, 52, 58, 60, 66, 127, 183, 192, 196

P

palpitations 16, 22, 48, 184, 190
pancreatitis 31
pancytopenia 8, 9
Pantothenic acid 112
papilloedema 98, 228
paraesthesia 23
parathormone 63, 194
parathyroid hormone 60
parotid 7, 29
Parotid Swelling 7
Paroxysmal nocturnal dyspnoea 16
Penicillamine 76
phosphate 65, 66, 68, 98, 104, 108, 119, 125, 126, 132
phosphorus 65, 66, 220
pickled foods 40
pneumothorax 16
polymyositis 28
Polyunsaturated Fatty Acid PUFA 129
pork 69, 104
potassium 38, 46, 157, 209, 220
potatoes 95, 113
poultry 40

Primary Osteoarthrosis 5
Prothrombin time 168
PseudoBarretts syndrome 155
psychogenic polydipsia 41
pulmonary embolism 16, 17
pulmonary oedema 16
Purging 47, 155
purpura 8

R

renal failure 8, 9, 56, 68, 143, 145, 146, 223
Renal osteodystrophy 145
rennin-angiotensin-aldosterone 216
Respiratory Acidosis 150
Respiratory Alkalosis 150
restlessness 23, 98, 184, 222
rhabdomyolysis 223
riboflavin 7, 88, 102, 103, 104

S

Scoliosis 24
Scurvy 113, 115, 116
seizures 41
selenium 38, 39, 83
septicaemia 17
Serum albumin 60, 61, 166, 167
Serum Amylase 171
Shellfish 73
sodium 27, 40, 41, 42, 44, 46, 59, 66, 132, 135, 153, 156, 205, 207, 215
sodium 38, 40, 41, 42, 50
soya milk 58
Spider naevi 8
spinach 52, 58, 95
spironolactone 209

Squat test 28
submandibular 29
syncope 16, 17, 22, 48, 61, 229

T

T3 178, 179, 180, 184
T4 178, 179, 183, 184
Tachycardia 20
tachypnoea 23
testosterone 7
tetralogy of Fallot 18
theophylline 46
thiamin 97, 98, 101, 108
Thiamine Deficiency 25
thiazide 47, 63, 208
thrombocytopenia 8, 9, 168
thrombopoietin 168
thyrotoxicosis 4, 20, 28, 34, 184, 185
thyrotoxicosis factitia 4, 20
thyroxine 83, 178, 181, 182, 183
Total Bilirubin 166
transferrin 79
tremors 28
TRH 179
triiodothyronine 83
troponin 20, 21
TSH 18, 178, 179, 180, 183, 184, 198

U

unconsciousness 23
urea 136, 137, 141, 142, 148, 157, 163, 228
Uric acid 157

V

vasovagal attack 16, 17
vertigo 26
Vitamin A 70, 87, 119, 120, 121, 122, 123
Vitamin B2 102
Vitamin B3 104
Vitamin B6 104, 108, 109
Vitamin B12 5, 74, 91, 93, 94, 95
Vitamin B complex 65, 103
Vitamin B compound 87
Vitamin B compounds 87
Vitamin C 8, 73, 87, 88, 113, 114, 115, 137, 187, 230
Vitamin D 60, 63, 66, 87, 124, 125, 126, 127, 195, 196
Vitamin E 87, 128, 129, 130
Vitamin K 8, 131, 132, 168
vomiting 6, 8, 29, 30, 31, 32, 41, 42, 44, 46, 47, 54, 72, 76, 80, 82, 91, 112, 121, 123, 145, 155, 184

W

Wernicke Korsakoff Syndrome 98, 101, 229
Wernicke's encephalopathy 97
Wet Beri-Beri 100
wheat 104
whole grain 52, 69
whole grains 52
Wilsons disease 76

Y

Yellow nails 5

Z

zinc 38, 39, 69, 70, 72, 82, 111
Zinc Gluconate 70

Printed in Great Britain
by Amazon